INTERACTIONS IN ARTIFICIAL INTELLIGENCE AND STATISTICAL METHODS

THE TECHNICAL PRESS — UNICOM
APPLIED INFORMATION TECHNOLOGY REPORTS SERIES

Editor: Gautam Mitra, Brunel University

Fourth Generation Languages and Application Generators
  Edited by David Martland, Simon Holloway, L. Bhabuta

Information Technology in Physical Distribution Management
  Edited by Robert L. Lewis

Expert Systems and Optimisation in Process Control
  Edited by Abe Mamdani and Janet Efstathiou

Computer Controlled Interactive Video: multi media authoring systems
  Edited by Tony Droar

Geometric Modelling and Computer Graphics: techniques and applications
  Edited by R. A. Earnshaw, R. D. Parslow, J. R. Woodwark

Effective Decision Support Systems
  Edited by John Hawgood and Patrick Humphreys

Software Quality Assurance, Reliability and Testing
  Edited by Chris Summers

Major Advances in Parallel Processing
  Edited by Chris Jesshope

Interactions in Artificial Intelligence and Statistical Methods
  Edited by Bob Phelps

Interactions in Artificial Intelligence and Statistical Methods

Edited by
Bob Phelps
BP Information Technology Research Unit

Technical Press

in association with Unicom Seminars Ltd

Published by Gower Technical Press Ltd, Gower House, Croft Road, Aldershot, Hants GU11 3HR, England

Gower Publishing Company, Old Post Road, Brookfield, Vermont 05036, U.S.A.

Based on papers delivered at a seminar on 16-17 December 1986 organized by Unicom Seminars Ltd.

ISBN 0 291 39743 3

British Library Cataloguing in Publication Data
Interactions in artificial intelligence and statistical methods.
— (The Technical Press — Unicom applied information
technology report series).
1. Mathematical statistics. 2. Artificial intelligence
I. Phelps, Bob
001.4'22   QA276

ISBN 0-291-39743-3

Printed in Great Britain by
Antony Rowe Ltd, Chippenham, Wiltshire

Contents

PART D: Statistical Methods in AI

Summaries

1. <u>The Application of Expert Systems in Statistics by D J Hand</u>

This paper explores the current state of the art of statistical expert systems. A close look is taken at what consultant statisticians do and what we might reasonably expect from statistical expert systems. Some systems built at the Institute of Psychiatry are described, including an interface to a multivariate analysis of variance package and two systems for helping to select appropriate statistical tools. The question of whether statisticians should regard the advent of the statistical expert system as a threat is discussed.

2. <u>Data Analysis as Search by David Lubinsky and Daryl Pregibon</u>

We equate data analysis with search. Since the search space is large and ill-defined, data analysts rely heavily on graphical displays to guide their search for good descriptions of a particular data set. We propose a more systematic study of the search space which ultimately reduces the dependence on graphics and lends itself to automation.

3. <u>AI and Generalized Linear Modelling: An Expert System for GLIM by J A Nelder</u>

GLIM is a statistical package useful for fitting generalized linear models. GLIMPSE is a front end for GLIM; it is written in Sigma-Prolog with APES as an expert-system shell. It includes a high-level command language to define tasks, and extensive syntactic help with task formulation. Analysis is structured in eight activities, and statistical expertise is coded as a sequence of tasks. The front end is transparent and libertarian. Future developments are outlined.

4. <u>An expert system approach for generating and testing statistical hypotheses by K M Wittkowski</u>

If common statistical analysis systems (BMDP, P-STAT, SPSS, SAS etc.) are used by scientists with little experience in the field of statistics, the wealth of methods currently available and missing protection against semantic errors often lead to erroneous application of methods and misinterpretation of results. Based on a discussion of typical errors, different sources of expertise, and several approaches in building expert systems, a new concept of structuring knowledge on statistical problems and methods is proposed. This concept is not only sufficient for a wide range of problems but also allows for relatively simple procedures of knowledge acquisition and application.

5. Dialogue Management with Computer-Based Statistical Analysis by S M Furner

It is argued that practical problems in the use of statistical software combines with the statistical knowledge of the user to influence the strategy employed within an analysis. Consequently, it is suggested that the integration of AI techniques into statistical software should address these practical issues, as well as provide a source of statistical expertise to the user. The resulting human interface would be an independent section of the statistical software package, containing various specialist intelligent systems co-operating to manage the interaction between the user and the application code.

6. A Computer Model with Expert Rules for the Control of African Cattle Diseases by G Gettinby

Trypanosomiasis and East Coast Fever are two important parasitic diseases of cattle that hold the key to improved livestock production in Central and East Africa. Since the early 1900s veterinary and zoological experts have reported their findings from studies on the transmission of the diseases. More recently contemporary experts in immunology, veterinary epidemiology and medicine have added to this knowledge base. However as yet no unification of the knowledge of past and present experts, experts working in different areas and in different organisations, has been undertaken with a view to quantifying the effect of control by chemotherapy, vaccination, etc. In collaboration with the International Laboratory for Research on Animal Diseases in Nairobi a first attempt has been made to construct a computer model for the study of the control of East Coast Fever.

7. AI and Stochastic Process Simulation by Ray J Paul

Complex stochastic processes can be modelled using simulation techniques. Formulating, controlling and interpreting such models are intelligent time consuming activities. Various attempts at using AI to make the modelling process more efficient will be outlined and discussed.

8. Intelligent Front End to Box Jenkins Forecasting by David P Reilly and Ana I Timberlake

ARIMA and Transfer function models are extremely useful for analysing and forecasting time series data. This paper discusses a software system - AUTOBOX (version 1.02) - which incorporates the three major steps in the model building - identification, estimation and forecasting - as a complete automatic feature, performed without any user intervention. AUTOBOX also offers the intervention detection techniques.

9. Learning_Tasks_Studied_in_Artificial_Intelligence_by Robert_C_Holte_and_Alan_J_MacDonald

Learning tasks are defined on the basis of (1) the type of information that is available to the learning system; and (2) the type of information or effect that the learning system is required to produce. This paper discusses the three basic learning tasks, and their major variations, that have been studied in Artificial Intelligence.

10. Learning_Diagnostic_Rules_from_Incomplete_and_Noisy Data_by_Ivan_Bratko_and_Igor_Kononenko

Methods for learning decision trees from examples provide good practical basis for automating knowledge-acquisition for expert systems. This paper describes the learning program ASSISTANT, developed from Quinlan's algorithm ID3 for induction of decision trees from examples. A number of improvements over the basic ID3 algorithm make ASSISTANT robust with respect to noise in the learning data, and applicable to problem domains where information is typically incomplete and unreliable. The effects of these improvements are illustrated by results of applying ASSISTANT to several medical diagnosis and prognosis problems.

11. Learning_If_Then_Rules_in_Noisy_Domains_by_P_Clark_and T_Niblett

Automatic rule induction systems for inducing classification rules from examples have already proved valuable as tools for assisting in the task of knowledge acquisition for expert systems. For the practical application of such induction systems, they should induce rules which both classify new examples accurately, and are easily comprehensible for the purposes of verification and explanation. In particular, they should be tolerant of noise and inadequacies of the description language used by the expert, as occur in many real-world induction problems. This paper presents a description and empirical evaluation of an induction system, CN2, designed with the aim of inducing short, simple, comprehensible rules in domains where problems of poor description language and/or noise may be present.

12. Synthesis_of_AI_and_Bayesian_Methods_in_Medical_Expert Systems_by_David_J_Spiegelhalter

Many proposals have been made for propagating the effects of evidence through a knowledge-base structured as a causal network. Ad-hoc numerical schemes are generally adapted but here we argue that it is both theoretically and pragmatically appropriate to stick to a strict probabilistic mechanism. Recent results in graphical representation of multivariate distributions and algorithms from the theory of relational data-bases are utilised in this framework.

13. <u>An Introduction to Statistical Pattern Recognition by B D Ripley</u>

In its early days AI meant pattern recognition, and pattern recognition techniques are now widely used for automated decision making. This paper introduces the basic idea of pattern recognition assuming no prior knowledge, and relates them to similar concepts within multivariate statistics. Even when the ideas are formally the same, the statistics and pattern recognition literature stress different aspects, especially computational speed in the latter.

Notes on Contributors

Ivan Bratko is professor at the Faculty of Electrical Engineering, E. Kardelj University, Ljubljana, Yugoslavia, where he teaches artificial intelligence, programming languages and computer data structures. He is head of the Artificial Intelligence Laboratory of The J. Stefan Institute, Ljubljana, and a visiting fellow of The Turing Institute, Glasgow. His major areas of interest are machine learning, qualitative modelling, AI-based robotics and applications of expert systems. Ivan Bratko is the author or co-author of over 90 technical publications, including a recent book, Prolog Programming for Artificial Intelligence (Addison-Wesley 1986).

Peter Clark studied for an MSc. in knowlege-based systems at Edinburgh University, 1984-5. Since then he has worked as a research scientist at the Turing Institute, Glasgow. His interests include the automatic generation of rules from examples, and the synthesis of plans from example sequences

Stephen Furner is currently a member of the Dialogue Engineering Group of the British Telecom Human Factors Division. He undertakes consultancy and experimentation into interactive user interface design to meet the needs of sponsoring units within British Telecom. He has been responsible for the design and analysis of human factors experimentation investigating aspects of user behaviour with communications systems. More recently this has tended to focus upon design issues for speech based interfaces. Mr Furner holds a C & G Full Tech. Cert. in Telecoms Engineering and a BA Hons. in Psychology and Sociology.

George Gettinby is a Senior Lecturer in Statistics at the University of Strathclyde. A consultant to International Laboratory for Research on Animal Diseases, Kenya and International Livestock Centre for Africa, Ethiopia. His interests are in modelling parasitic diseases, forensic science, clinical trials and micro-computer software for data handling.

David J Hand is a Senior Lecturer in Statistics at the London University Institute of Psychiatry. He is also a Director of Sigma X, the statistical consultancy. His research interests include pattern recognition, multivariate analysis, and statistical expert systems, the latter a field in which he has been working for several years.

Robert C Holte graduated with an M.Sc in Computer Science
from the Univerity of Manitoba in 1977. He is now in the
final year of study towards a Ph.D in Machine Learning.
Currently he is a research assistant in the Department of
Electrical Engineering and Electronics at Brunel University.
He holds a two year grant from the Commission of the
European Communities Cost - 13 project on Knowledge
Acquisition and Machine Learning.

Igor Kononenko lectures at the Faculty of Electrical
Engineering, Ljubljana, Yugoslavia, in artificial
intelligence and programming languages. His major research
area is machine learning.

David Lubinsky is a Consultant to the Statistical Computing
and Data Analysis Research Department at AT & T Bell
Laboratories. He is currently enrolled in the Ph.D Program
in Computer Science at Rutgers State University,
New Brunswick, NJ. His area of specialization is Artificial
Intelligence with further specialization in Machine
Learning.

Alan J MacDonald graduated with B.Sc in Electrical
Engineering from Glasgow in 1977. From 1977 to 1979 he was
a Design Engineer at Racal Datacom. He received his Ph.D
degree in Machine Learning at Brunel University in 1984. He
held an SERC IT fellowship at the Department of Computing,
Imperial College. Currently, he is a lecturer in the
Department of Electrical Engineering and Electronics at
Brunel University.

John A Nelder is visiting Professor at Imperial College. He
was Head of the Statistics Department at Rothamsted 1968-84.
He originated the statistical packages GLIM and Genstat,
which incorporate the ideas of generalized linear models.
He is a Fellow of the Royal Society and past President of
the International Biometric Society and the Royal
Statistical Society.

Tim Niblett received a PhD. in machine intelligence from
Edinburgh University. After post-doctoral work at the
Universities of Illinois and Edinburgh, he became one of the
founding members of the Turing Institute in 1983. Since
1984 he has worked jointly at the Turing Institute and the
University of Strathclyde. His interests are in machine
learning and expert systems.

Ray J Paul took a first degree in mathematics, and a masters
degree and Ph.D. in operational research at Hull University
prior to joining the LSE as a lecturer in operational
research. His research interests have included stock
control and production scheduling, on which he has published
several papers, but recently have been increasingly
concentrated in simulation. Consulting in defence and
industry have formed the basis of a variety of software
developments in simulation, including artificial
intelligence systems.

Bob Phelps has a background in the mathematics of decision making and has taught statistics and operational research at Brunel University. He has worked in statistics, O.R. and A.I. and has particular interest in the interaction of these disciplines, about which he has written, spoken and organised a conference stream. He currently work with BP's Information Technology Research Unit.

Daryl Pregibon is Supervisor of Statistical Computing and Data Analysis Research at AT & T Bell Laboratories at Murray Hill, New Jersey. He received his Ph.D. in Statistics from the University of Toronto in 1979. Dr Pregibon spent one post-doctoral year at Princeton University and another at the University of Washington. He has been at AT & T Bell Laboratories for the past five years. Dr Pregibon has been involved in the development of knowledge-based software for statistical data analysis since his coming to Bell Laboratories. He is a co-developer, with W A Gale, of first prototype expert system for data analysis, REX.

David P Reilly is the founder of Automatic Forecasting Systems, Inc. and a lecturer at Drexel and Penn State Universities in the US. He has over 20 years experience in the use of time series techniques, specifically the Box-Jenkins methods and has developed the AUTOBOX algorithms. AUTOBOX has been available on the Computer Sciences and Chase Econometrics bureaux for 10 years.

Brian D Ripley is Professor of Statistics at the University of Strathclyde, having previously been at Imperial College, London. His major areas of research are spatial patterns and processes, simulation, image analysis and pattern recognition.

David J Spiegelhalter is a statistician in the Medical Research Council Biostatistics Unit in Cambridge. Much of his work involves the development and evaluation of aids to clinicians making difficult decisions, and this has naturally led to an attempt to synthesise AI and statistical techniques.

Ana I Timberlake is the Managing Director and a Senior Consultant of Timberlake Clark Ltd, a management consultancy specialising in statistical and economic applications and software. Mrs Timberlake has over 20 years of consulting experience in the area of statistical analysis.

Knut M Wittkowski is Biostatistician and Assistant Professor of Biomathematics at the University of Tubingen (F.R. Germany). He has also held positions at Siemens AG, Berlin (data base applications) and University of Gottingen (computational statistics). He received a Ph.D degree in mathematics and informatics at the University of Stuttgart for the development of a statistical expert system. Dr Wittkowski is author/co-author of several papers on representation of statistical knowledge, man-computer-

interfaces and nonparametric statistics. He has taught
medical biometry and statistical computing at Gottingen and
Tubingen University, and he is a member of the Biometric
Society and the IASC. His main research interests are in
the application of new statistical strategies and
AI-techniques to the design of statistical expert systems.

Editor's Preface

This book brings together papers dealing with an exciting new area of research and application which is the result of interplay between two traditions of problem-solving, artificial intelligence (A.I.) and Statistics. Interaction between two such traditions offers the possibility of revitalization by bringing fresh perspectives to bear on old problems. Within any tradition problems, techniques and methodologies tend to become established as a culture, in Kuhn's phrase a "normal science", and this can lead to an inward-looking, "ivory tower" mode of thought. A cultural establishment of this sort can be revitalized by a "cultural revolution" such as happens occasionally within subjects: in statistics the introduction of Bayesian methods was one such landmark; the influence of A.I. techniques is another.

In contrast to the scientific analogy whose new theories may actively displace old ones, the nature of both A.I. and Statistics encourage the gradual accumulation of knowledge, by integrating the best of previous ideas with the best of the new. When the integration is of ideas from another discipline, it is essential that respect for that discipline is first established. The high level of "hype" surrounding A.I. has on occasion encouraged an "emperor's clothes" view of A.I. and a suspicion that programs are written as an end in themselves rather than as tests of underlying principles or as embodiments of useful results. On the other hand, the complexity of many statistical methods has resulted in their being discounted as irrelevant to human thought processes, insatiable in their requirements for numerical data and computationally intractable in all but artificially simplified domains. Fortunately, these initial suspicions are succumbing to a new spirit of co-operation: this book bears witness to that new spirit.

The first section of this book "Automating Statistical Methods" deals with the representation and use of knowledge which statisticians employ in using statistical techniques. This knowledge, suitably coded as an expert system interfacing with the statistical technique, forms an environment for the user which guides the use of the technique and helps in interpreting the results.

The provision of such an interface not only makes use of statistical techniques quicker and easier, but also helps prevent misuse of the techniques by ensuring checks on the validity of the method for the data and providing an accurate statement of inferences which can be drawn as a result of the analysis.

David Hand discusses these issues and possible structures for representing expert statistical knowledge; he illustrates his theme by reference to several systems he has developed for use in multivariate analysis of variance, the selection of tests and non-parametric statistics. David Lubinsky and Daryl Pregibon describe their TESS system in the context of regression analysis where they introduce concepts of interestingness and accuracy, applied to different ways of describing the data. They pay particular attention to the importance of problem context, which is difficult to include in an automated system, and argue for multiple investigations using different data descriptions to be evaluated by the system user. John Nelder writes about the GLIMPSE front end for the well known GLIM linear modelling package. He stresses the "libertarian" philosophy that GLIMPSE embodies which entails the user interacting with the system, the latter offering advice but not dictating the course of the analysis. The range of features available in GLIMPSE is surveyed. Knut Wittkowski deals with the selection of statistical techniques for a given data set. He proposes a pattern matching approach, based on a classification of statistical problems in terms of knowledge about the domain and the data. His claim is that the information specified by this classification is sufficient to ensure the choice of an appropriate statistical technique.

The second section "Integrating A.I. and Stochastic Modelling", looks at two systems successfully combining aspects of A.I. with statistical models. George Gettinby describes an application to Africian cattle diseases, where an epidemiological model is comprised of statistical simulations the use of which is governed by expert derived rules. The combination of these techniques allows the construction of a large scale, modularly structured model. Ray Paul is concerned with automating the process of constructing a simulation from a linguistic description. He describes the AUTOSIM system which generates code from a given specification, and describes work on a natural language understanding system which accepts text descriptions of the simulation. Together, these components leave the analyst free to concentrate on understanding and formulating the problem.

The third section, "A.I. Approaches to Learning from Data", sets out some of the A.I. work which has tackled the problem of finding structure in data sets, more especially of finding rules which describe the data well enough for classifications of data points to be achieved. To prepare the ground, Rob Holte and Alan McDonald survey the field of learning in A.I. They describe the types of problem that have been tackled, explain the terminology used and indicate problems that are still to be overcome. Ivan Bratko and Igor Kononenko set out the ASSISTANT algorithm which produces a tree structure of rules to perform classification of data from an initial set of given classified examples. To deal with natural variation (noise) in the data, they employ a pruning mechanism on this tree. The method is evaluated on sets of medical data with encouraging results, as well as giving simple sets of easily understood rules. Peter Clark and Tim Niblett are also concerned with producing rules to discriminate between classes of data, but start with an A.I. "conceptual clustering"

algorithm which produces rules to characterize individual classes
rather than a discrimination rule tree. They are also concerned
with the problem of noisy data and introduce a statistically
influenced criterion for limiting the number of rules used.
Again, evaluation on medical data is discussed.

The fourth and final section, "Statistical Methods in A.I.",
explores examples of the use of statistical techniques applied to
problems in A.I. The characterization of uncertainty has a
chequered history in A.I., with initial enthusiasm for
probability being replaced by new approaches ranging from the
ad-hoc (uncertainty factors) to the formally complex
(Dempster - Shafter) and from numerical measures to linguistic
descriptions. A major reason for probability not generally being
embraced by the A.I. community is the apparent complexity of
conditional distributions necessary for all but the simplest
problems. David Spiegelhalter proposes a method for propogating
probabilities accurately through a knowledge base considered as a
causal network (a standard A.I. representation). This involves a
relatively low number of probability assignments and allows a
sound theoretical basis for the treatment of uncertainty in
knowledge based systems. Brian Ripley surveys the field of
statistical pattern recognition. Pattern recognition is an
innate perceptual ability of animals carried out with apparent
ease, and yet it is an area that has proved remarkably resistant
to A.I. investigations. It is also an area where the immediate
applicability of statistical techniques is most marked.
Brian Ripley distinguishes between the structural and statistical
aspects of pattern recognition and discusses the importance and
the limitations of the statistical approach.

An interesting feature of the conference on which this book is
based, was the question, repeatedly asked by the audience, as to
why the speakers felt that what they were doing differed from
traditional statistics or traditional A.I. The fact that such a
question can be asked shows in itself that both subjects have
acquired wider boundaries, and that ideas absorbed from each
other are beginning to win acceptance as a normal part of their
respective cultures. It is to be hoped that this diffusion of
ideas will continue and that workers in both disciplines will
come to regard their counterparts on essentially studying the
same problem from different points of view: the problem of
finding structure in the world and using it to guide decisions.

PART A
Automating Statistical Methods

1 The Application of Expert Systems in Statistics

D. J. Hand
London University Institute of Psychiatry and Sigma X

1. INTRODUCTION

The numbers of people who rely on statistical results and techniques is growing. In addition, computers are now very widespread, giving people access to numerical power and statistical software on an unprecedented scale. And yet it is clear that the level of statistical expertise in this rapidly growing user community is not growing at the same rate. There are dangers in this. Hooke [15] says: "Use (of statistics) has been replaced by overuse and misuse. Regression is being used in foolish ways in the neighbourhood of almost every computer installation," and one can easily find many other statements in the same vein. Recent publications [5,23] have criticised the level of statistical expertise demonstrated in medical journals. It seems that there is an increasing need for statistical education of non-statistician users of computer software and for the training of many more statisticans. And yet this is unlikely to happen.

This problem is not unique to statistics. It is simply a symptom of the increasingly complex and information oriented world in which we live.

One possible answer, if one cannot train more experts and if consumers of statistical results do not have the time to improve their grasp of the underlying methodology, lies in building intelligent computer programs which can act, at least in some measure, in the role of statistical experts. Such a program would be a statistical expert system.

Thus a very broad definition of a statistical expert system would be that it is a computer program which interacts with users to guide them in some aspect of analysing data, helping them to choose and correctly apply appropriate statistical techniques. This definition suggests that they play the role of experts (which is implicit in their name) but it is so broad that it obscures rather than illuminates precisely what it is one might expect such systems to do. We begin to examine the situation more closely by addressing the question of what one might reasonably expect a statistical expert system coming on the market in the next few years to do.

First we should note that statistics is a big subject.

- It's big in the amount of knowledge that it encompasses (consider the number of books and papers that exist on statistical methodology).

- It's big in the types of things that a statistician is expected to do. Thisted [21] describes the complete expertise of an expert data analyst as encompassing such areas as 'mathematical statistics; techniques of graphical display and analysis; rules of thumb for judging the importance of apparent indications; copious examples of bad or misleading analyses (coupled with a catalogue of common errors made by novices, the avoidance of which is essential to respectability); methods, both ad hoc and those thoroughly grounded in theory, for basic operations such as smoothing, assessment of variability, and model building; and - perhaps most important - knowledge of how and when to elicit specific subject matter information from a scientific collaborator." One might add that a familiarity with several statistical computer packages is also essential.

Because of this size, whatever one might hope that statistical expert systems will do eventually, it is clearly unrealistic to expect them to do everything an ideal expert statistician might do at present. We will have to wait a few decades for that. However, it would be reasonable to focus on some subdomains of statistical consultancy practice so that we can develop systems for those restricted domains.

Focussing on subdomains, however, has its attendant problems. Consider, for example, the 'cookbook effect' - the belief that statistics is composed of isolated techniques, possessing little underlying similarity, and that the major aim of a statistician is to choose the right method to plug the data into. This misunderstanding of what statistics is all about is pernicious. It must be one of the reasons underlying the bad press that statistics sometimes attracts (the more general reason being, of course, that criticism which should more fairly be directed at ill-formulated research questions or poorly designed studies in fact is directed at the statistical tools used to try to remedy these pre-existing inadequacies).

So that we can choose a suitable subdomain of statistical expertise for encapsulation inside an expert system, we must look closely at the sort of things that statisticians do so that we can see what we might expect of our statistical sorceror's apprentice (and I use this term deliberately, with its connotations of something that could get out of control. We'll come back to the issue of whether statistical expert systems should be seen as a threat later on). 'Suitable' here must mean that it is possible to code it in some computer programming language, that it is possible to do this in a reasonable time, and that the result will be of some value to someone. A system which was too highly specialised would never be used. Existing packages like SPSS, SAS, GLIM, GENSTAT, and BMDP are successful by very virtue of their generality.

2. WHAT THEN DO STATISTICIANS DO?

The tasks below are to be seen as different things that a
statistician does, and not as a sequence of things done. More
often the statistician will do all of them several times
during the course of any given research project. Occasionally
only one of them will be required.

2.1. Refine research objectives

Statisticians interact with clients to refine the research
objectives. That is, they discuss operational definitions of
the things the researchers are interested in, they discuss
possible sources of data, they explain principles of
experimental design, they perform back-of-the envelope
calculations to explore sample sizes.

2.2. Identify relevant techniques

While discussing the aims of the researcher, they'll be
thinking about the sorts of techniques which might be
relevant. I think any statistician involved in designing a
research project will have some idea of what sort of
statistical techniques he will use to analyse the results.
This is, unfortunately not always the case with
non-statistician researchers. Many consultant statisticians
have bemoaned the tendency for researchers to collect their
data before seeking statistical advice [13]. Certainly I have
far too much experience of researchers coming to me with their
data and asking how to analyse it. In such cases one goes
through much the same sort of exercise as in the ideal case,
discussing the researcher's aims and objectives, and
attempting to cast these aims into the formal mould of some
class of statistical techniques. The danger now is that the
aims may defy such formalisation. In this case the
understanding consultant, in order to rescue the researcher's
PhD or whatever, attempts to help the researcher to think of
some relevant question which can be tackled using the
available data!

2.3. Analyse the data

This consists of examining the data which has been collected
and applying one's chosen techniques to it. This will usually
involve choosing and using some computer package. Typically,
many techniques will be used, and this is what distinguishes
between real life statistical practice and the idealised world
of the text book. In 2.2 above we chose the method. Here we
apply it. But in fact we find that the application involves an
iterative and cyclic approach. Transformations must be applied
to resolve problems of nonlinearity or heteroscedasticity.
Doubts are raised about the wisdom of treating merely ordinal
variables as interval. Effective ways have to be found to
summarise large numbers of variables before patterns of
relationships between different classes of variable can be
explored. Missing values raise questions of the relevance of
the technique one had originally intended using. Or one

discovers that the researcher omitted some vital piece of information - the groups comprised the same people and were not independent after all. Or one just thinks of a better way to do the analysis.

The analysis of data is not something which is done in the dark. It is done in conjunction with the researcher. In general the statistician will not be an expert in the domain area as well as in statistics (there just is not time to acquire both classes of expertise - see [8]). It will be necessary to discuss with the researcher the assumptions that the methods make. For example, such-and-such a method assumes Poisson distributions, is this a sensible assumption? (And, in the vein of the paragraph above, can the assumption be tested using the available data ?). Or as another example, suppose one discovers an interaction between two factors. How best should one attempt to examine this? Does it make sense to split on one of the factors rather than another? This is a question which should be directed to the substantive expert - the researcher - rather than the statistician.

2.4. Discuss the conclusions with the researcher

By definition the researchers are not statisticians. That is why they sought expert statistical help in the first place. This means that it is no good presenting them with the computer output of the principal components analysis or the manova run. Some explanation and discussion will be necessary. This level of the statistician's work will almost certainly lead back to one of the earlier levels - except in the rare cases when the question being posed was a very simple one indeed.

We thus have four distinct types of activity undertaken by the statistician. However, they should not be seen as stages - it is very rare that one can simply proceed from the first through to the last and then finish. Instead, the exercise of statistical consultancy is perhaps best represented as a tangled web!

We have here something of a dilemma. A proper view of statistical consultancy must contain a very wide range of skills but to attempt to build a computer program containing all those skills would clearly be a mammoth undertaking. If one sacrifices generality for practicability then one is at risk of promoting the cookbook fallacy and one runs the risk that the resulting product will be of such limited scope that it will be of virtually no practical value or interest.

However, we can distinguish several different types of activity that a statistical expert system might usefully do:

 - help in choosing relevant and appropriate techniques. Systems aimed at this include [6,9,17].

 - help the user to apply a chosen tool correctly. That is, the system embodies a **strategy** for some technique and guides the user through the application of the strategy. Systems which do this include REX [4,19], the programs developed for regression analysis and survey analysis at the University of Canterbury (eg. [22]), and Student [3]. This latter is an ambitious attempt to develop a system which itself builds expert systems for guiding the user through the strategy of any specified technique. Statistical strategy is itself becoming a topic of research interest, having implications for methodological research and statistical education, as well as expert systems development (see, for example, [12,18]).

- interface the user to an existing statistical package. Examples of this include BUMP [20] and GLIMPSE, the GLIM interface being developed by John Nelder.

We keep referring to the user. Obviously this person plays a central role. Design attributes will clearly be critically affected by the characteristic of the user. So let us examine potential users in a little detail.

3. WHO WILL USE THE SYSTEM?

Most work on statistical expert systems to the present date [11] has implicitly assumed that the user will be statistically naive. This is in contrast to most of the work on medical expert systems which has assumed that an expert - or at least someone who understands the technical jargon - will be using the system, although it is also an assumption made in some systems currently under development for other domains.

The assumption that the user will be statistically naive has implications for the way the system will be exepcted to work. Hand [7] lists the attributes systems intended for use by such researchers should have. They include the facility to proceed in small steps, with much explanation, that the system should drive the interaction, that it should be able to demonstrate examples of statistical phenomena and concepts, and, of course, that there should be a very comprehensive and flexible capacity for answering the user's questions - about technical terms and about why it, the system, wishes to know something about the data. This might involve giving simple illustrations of statistical phenomena.

Of course, it is important to recognise that there is not just one 'statistically naive' class of user, all the members of which have the same level of knowledge of statistical concepts. Instead there is a continuum. To be of any real value the system must be flexible. One saw the beginnings of this in systems such as the SCSS Conversational System [16], which could operate 'in three prompting styles: verbose, normal, or terse. The user is free to choose among them and can vary the prompting style at any point in the session. Normal prompts ask a complete question and give suggested responses or helpful information. Terse prompts ask abbreviated questions (a maximum of eight characters).' A more recent illustration of such flexibility is given in Nelder's GLIM interface, GLIMPSE.

One exception to the assumption that the user should be relatively unversed in the art of statistics is our most recent system. This system does not act as an expert - which we feel would be redundant since it is being used by an expert. Instead it acts with the expert to (we believe) improve the quality of statistical advice being given. In part this system has its genesis in the feeling that one could give better statistical advice if only one had more time to devote to each study. This is a reference to one consequence of the growing demand for statisticians referred to at the start of this article.

4. SOME SYSTEMS BUILT AT THE INSTITUTE OF PSYCHIATRY

The Institute of Psychiatry is a postgraduate medical school of London University. Around 400 researchers from a variety of disciplines work there, including psychologists, psychiatrists, biochemists, pharmacologists, neurologists, and statisticians. My job, in conjunction with the other statisticians in the Biometrics Unit, is to give statistical advice to these researchers. This includes advising on experimental and survey design, advising on the analysis of data, and advising on how to use various statistical packages (including SPSSX, SAS, GLIM, GENSTAT, etc). Sometimes, of course, clients come in with problems which require one to explore an area of statistics with which one does not have a great familiarity. Sometimes it is even necessary to develop entirely original approaches.

It will be clear that the demands on the time of the statisticians are high, and it is because of this that I originally became interested in statistical expert systems.

We have explored several different types of expert system. Different in that they are aimed at different statistical tasks from amongst the range of tasks described above. Different in that they use different internal architectures to achieve their aims.

In some cases the systems never reached the stage of being used in statistical consultancy practice. In others they were tested both by statisticians and non-statistical researchers at the Institute. In both cases we learnt a good deal about how such systems should be designed and what one should aim to achieve with them.

In this section we summarise our work.

4.1 BUMP

BUMP [20] is an acronym for Biometrics Unit MULTIVARIANCE Program. BUMP is an interface to MULTIVARIANCE. MULTIVARIANCE is a program for generalised univariate and multivariate analysis of variance, covariance, regression, and repeated measures analysis. Such techniques are heavily used at the Institute, especially by psychophysiologists. MULTIVARIANCE is not easy to use. First there is the problem of the basic complexity of multivariate analysis of variance. There are relatively few non-mathematical guides to the technique and the researchers we encounter cannot afford the time to acquire sufficient ease with the necessary mathematics. (We hope to have slightly eased the situation in [14]). Second, MULTIVARIANCE uses a system of numerical codes to describe the analysis to be performed. (This has also been eased by MULTISTAT [2], a preprocessor which uses mnemonic keywords. This, however, was not available to us at the time we built BUMP). The difficulties mean that researchers often end up using less powerful programs which may not suit the data so well - perhaps requiring unjustified and unnecessary assumptions. Note that the difficulty of using MULTIVARIANCE means that it will be seldom misused - one has to have a good idea of what one is about before attempting to do it. But, at least in this case, preventing misuse of this particular program does not resolve things if researchers then go on to mis-apply some other technique: it merely leads to some other kind of abuse.

We should, perhaps, here note a contrast with SPSSX's MANOVA program. This is very easy to use in that it is part of the SPSSX system, with the usual syntax of keywords. However, in our experience it is also easy to misuse. We have often witnessed researchers with an inadequate grasp of multivariate analysis of variance misspecifying their research design or objectives. The two programs, MULTIVARIANCE and MANOVA, provide examples of opposite problems, but one would hesitate to say which was worse. We should also add that the attractiveness of general packages, referred to above, has meant that nowadays researchers at the Institute tend to use SPSSX's MANOVA and not MULTIVARIANCE.

There are several advantages of building an interface to an existing package, compared to building an 'intelligent' program to do the kind of analysis in question from scratch. For example, one does not have to worry about the numerical manipulations. These are taken care of by the target package. Similarly, the target program might have highly specified forms of syntax, so that there is less freedom in the final output format of the interface. This was the case with MULTIVARIANCE, which is intended for batch use. There might also be disadvantages, of course. There might be a problem interfacing the conversational part of the program with the numerical side. (REX, [4,19] and other systems functioning under UNIX, of course, do not have this problem).

BUMP was our first foray into the 'expert system' arena and was deliberately limited in scope. It functioned as follows.

The researcher first discussed the design, objectives, and data with a statistician. (We are assuming here that the statistician in question had not seen the project before. A situation unfortunately all too common). This served to familiarise the researcher with any necessary statistical terminology (though BUMP required relatively little - see below) and helped to clarify precisely what questions might usefully be explored using MULTIVARIANCE. The researcher then responded to an interrogation by BUMP, through which the latter built up an internal representation of the design and objectives. Sometimes deduction was needed to fill in any gaps in supplied information, and occasionally defaults were adopted. From this BUMP built a MULTIVARIANCE program and combined it with the data. BUMP ran on our local mini-computer whereas MULTIVARIANCE ran on the mainframe at ULCC. BUMP gave the user the option of entering the data by hand, or of specifying a file name containing the data (either a local file or a remote file). When all necessary information had been supplied, BUMP dispatched the completed program for processing.

BUMP had HELP frames, which gave additional (canned) advice on anything the user was uncertain of. It used a conventional programming architecture, rather than, say, production systems, because we felt that this was more in accordance with its aims. In particular, we did not have anything which matched the diagnostic paradigm implicit to production systems. Instead we were presented with the problem of constructing an internal representation of a precise structure.

But BUMP was limited. As an interface it could be severely criticised on the grounds that it only tackled one half of the problem. Its job was finished once the program was dispatched, and it did nothing to help explain the results to the user. Moreover, it did not interact with the data as well as the user - a feature of more advanced statstical expert systems.

Despite these limitations BUMP can be regarded as a success. It attracted considerable interest from researchers at the Institute and was successfully used.

BUMP took about twelve man months to construct, including the HELP file. It was written in FORTRAN and occupied about 1.5Mb

4.2 STAT1

Our second program tackled an entirely different area. Potential users of multivariate analysis of variance are necessarily involved in the larger projects. They might be PhD projects or long term team research projects extending over several years. Another class of our client is the very short term - an MSc student or a junior medical doctor attempting to do some part-time research for example.

These kinds of cases often require one to identify an appropriate test, summary statistic, appropriate measure of correlation, etc. There is little problem here in applying the method, once chosen, or in running the computer (or calculator) program. The problem lies in identifying the appropriate method. Of course, for any one project the identification is trivial. The problem arises because of the large number of such projects and the diversity of statistical techniques that they require.

Our second program (STAT1 - we abandoned the attempt to think of interesting acronyms once we recognised that we would be building a number of such programs) was aimed at helping the statistically naive researcher identify an appropriate tool [9]. One of the things we have been particularly struck by in our exploration of statistical expert systems is how very domain dependent the choice of method is. It has often been remarked that statistical consultants can only perform well within one specialised area - outside that they do not have sufficient background knowledge to be very effective. Not only will there be emphasis on different methods and classes of method, but even the 'same' method will be used in different ways. REX, the Bell Labs regression program is oriented towards a way of doing regression analysis which does not seem to address the purposes of most of our researchers. It arises, presumably, because of the difference between engineering and psychosocial applications of the technique. In any case, STAT1 was aimed at helping social science researchers choose appropriate methods.

Note that while one might criticise programs with this objective, one should also note that they could be especially beneficial. They, one hopes, will be immune from the human tendency to force all research questions into the mould of one's favourite statistical technique.

STAT1 was intended to act as a baseline against which other programs with the same objective could be compared. It was not intended to be used outside the Institute, but was meant to gauge user reaction and to help us to identify any particular areas of weakness in the general approach.

An obvious choice of structure for a system to help choose techniques is the decision tree. Andrews et al [1] have presented an impressive static display, in book form, of just such a tree. We took their tree, extended it slightly, and converted it into a computer program.

The primary knowledge structure of the program is a file of frames. Each frame consists of a question, a menu of possible responses, and control information about which frame should be displayed following each possible response. Only the question and list of possible responses are displayed to the user. In all, the tree has 246 such frame nodes.

Access to the frame file is via a frame directory, a file containing details of where in the main frame file each frame is located.

Since each frame contains details of its successor nodes the executive program is very simple. The 'data' itself provides the control structure. As well as continuing on down the tree (if one gives one of the answers specified in the menu) the user has the options of backing up one node (from where one could take a different path down the tree) or of returning to the root of the tree.

In parallel with the frame directory and frame file there is a help directory and help file. This latter contains canned explanations about technical terms. This is accessed by the user being invited to type 'Xword' for 'explain word'. The only difference between this and the frame-file/frame-directory structure is that this can access itself whereas the frame structure must always loop through the executive.

During operation, invisibly to the user, STAT1 records a trace of the frames accessed. This allows us to assess the performance of the system (do any nodes in particular cause misunderstandings?) and to keep a statistical summary of what sorts of techniques are most heavily used.

Weaknesses of decision trees are well known. Most of them are a product of the basic inflexibility of the structure. Thus, they are difficult to modify consistently and they do not handle 'don't know' responses easily. They are restricted to requesting their information in a fixed order, though they can easily make use of any unsolicited information the user may supply (by keeping this information in a separate blackboard).

One weakness of this particular program which our users identified was an inadequacy in its power for explaining things. We have taken note of this and in the next section describe a structured production system approach to technique selection which uses an entirely different explanation method.

4.3 STAT2

Another obvious choice for a system for aiding in selecting technique is the production system. Straightforward pure production systems proved inadequate for our aims, and we explored a structured system. First, however, a word about the power of production systems for deep 'deductive' chaining through rules. We believe that, certainly as far as statistics is concerned and perhaps for other disciplines as well, this facility is seductive and misleading. For choice of statistical technique breadth is far more important than depth. When an expert is asked why he chose a particular method he does not say 'I chose A because B and C were true. And I knew B was true because D was true. And D was true because ...' Instead he identifies the major points of the technique and simply lists them. Further questioning produces other, less important points. And so on. A 'major' point here is probably one which distinguishes the technique from many others, rather than few others. (For example, 'because I wanted a nonparametric method').

This has prompted our structure. We have a broad bottom layer of rules containing definitions of statistical methods. The antecedents of the rules constitute the definition: 'if A and B and C ... then the method being described is regression analysis'. The antecedents are grouped into two sections, the first being a description of the aims that the technique seeks to achieve and the second a description of the conditions which must be satisfied before the technique can justifiably be used. The antecedents here are couched in a statistician's language - they make free and explicit use of technical jargon.

These base level rules are ordered according to the probability that each technique will be needed. Thus more commonly used techniques come near the beginning of the list. This ordering can, of course, be different for different working environments. In use the system first explores the antecedents of the first rule at this level. Any unsatisfied antecedent causes the rule (and associated method) to be rejected. When this happens the remaining rules are reordered according to the degree to which their antecedents are satisfied by the information so far collected.

This bottom layer of rules is fed by a production system which reformulates the technical statistical terms in simpler language. Questions may be broken down into simpler sub-questions, alternative words might be used, and so on. In general, the further back up the implicit tree of the production system one goes the less technical the formulation becomes. This approach has several merits. In particular, it allows the system to adapt itself to the level of expertise of the user. Initially the system opens its exploration of the user's aims at the bottom level of rules (the 'high-level' statistical concepts). Someone who was familiar with the terminology would directly answer the questions and the system would match the responses directly against the antecedents of the base level's rules. Someone who was not familiar with the terms used would be unable to respond and the system would reformulate them in simpler, albeit more extensive, language.

One can think of this approach as being an attempt to replace the 'canned definition' approach to providing help by something more flexible.

One further aspect of the system is worth mentioning. When a base level rule is satisfied, the user is informed that the right hand side of this rule is one possible appropriate technique. However, the system does not stop there. It continues to explore the base level rules to see if any of the remaining unrejected ones also satisfy the requirements. The user is told that the system is searching for even better methods. Should more than one base level rule fire then the final ordering of recommendation will be according to the number of antecedent conditions in each rule - the harder something is to satisfy, the more specific it is likely to be and so the more powerful a technique it is likely to be for the particular objectives of the researcher.

It goes without saying that this is a mixed initiative system. Any unsolicited information that the user provides about the task will also be retained in the production system's blackboard and will be used by any rules which can make use of it.

STAT2 is described in [9].

4.4 KENS

KENS is a system being developed completely separately from my work at the Institute of Psychiatry. KENS is a Knowledge

Enhancement system for Nonparametric Statistics. Of the systems described so far it is the most advanced, and we hope to distribute it.

If one looks through sections 4.1 to 4.3, which are in the correct time sequence of development, one can see a trend away from the statistically naive researcher towards the expert statistician. This reflects the growth of my own conviction that what I need is a system that can help me do my job of statistical consultant better, rather than a system that can replace me in doing particular parts of my job.

I should come clean and state that KENS is not an expert system in the conventional sense. An expert system acts **as** an expert, as we said at the start. KENS, in contrast, acts **with** an expert and (in principle) helps the expert to do an even better job.

KENS has several motivating influences, including a recognised inadequacy in the way classic expert systems approach problems, the fact that it is easier for people to recognise things than remember them, and the information explosion.

5. THE FUTURE

Perhaps the first question that occurs to many consultant statisticians when they encounter statistical expert systems is whether or not the systems will pose a threat. They are, after all, intended to act as consultant statisticians. As far as I know, no similar anxiety has been widely debated regarding the use of medical expert systems. The difference between the two domains lies in the likely intended user. In the former case it is the statistician's client who may well be using the system directly while in the latter case it is an expert, who may merely be seeking a second opinion. Is the statistician therefore doomed to become redundant?

The immediate answer to the question is, of course, no. If statistical expert systems do pose a threat then it will not become apparent for many years. Instead, such systems seem to present an opportunity: they can free consultant statisticians from the particularly mundane so that they can concentrate on the more challenging and interesting aspects of their work. This was the idea behind BUMP, which freed the statistician from having actually to code a MULTIVARIANCE program, a relatively tedious and mechanical task, so that he could spend more time exploring design considerations and the researcher's aims.

However, before we simply dismiss the notion that statistical expert systems might be a threat, there is another aspect which we must at least address. This is that such systems might be a threat to **statistics** as a discipline. Might they, for example, lock in certain approaches to statistical practice, might the advent of poor systems have a deleterious effect on the image of the discipline, and might the existence of default decisions (in strategy systems, for example) impose patterns on data rather than discovering patterns in data? On the last point, are we in fact in danger of automatic data analysis? And, connected with this, how insistent should a system be that a user follow its recommendations? If the system says that a particular method is most appropriate can the user insist that the system uses another method or, if the system says that the non-linearity of the response variable mandates a transformation, is the user able to force the system to analyse the data without making a transformation?

There is an associated ethical question, perhaps less pressing than the analogous question in medical applications of expert systems ([10]), as to who is responsible if a mistake occurs through the use of a statistical expert system. Is it the user or the programmer?

The general answer seems to be that there are dangers. But perhaps no more than those already presented by existing software: already there exist poor packages which promote bad analyses, already there are hints of automatic data analysis in such things as stepwise selection of variables in regression analysis, already if an existing package contains a bug which leads to a mis-analysis of a clinical trial or inappropriate financial investment there are ethical (and, perhaps, legal) problems.

Perhaps we should just remark that if the history of statistical packages to date is anything to go by, it seems that there will be many expert systems, some good, some bad.

What is important, though, is to try to make sure that as many as possible are good. This means that it is vitally important for statisticians to take an interest in such systems.

REFERENCES

1. F.M.Andrews, L.Klem, T.N.Davidson, P.M.O'Malley, and W.L.Rodgers, A guide for selecting statistical techniques for analysing social science data, Survey Research Centre, Institute for Social Research, University of Michigan (1981).

2. J.D.Finn, MULTISTAT/MULTIVARIANCE Manual, National Educational Resources Inc., Chicago (1981).

3. W.A.Gale, Student phase 1 - a report on work in progress. In 'Artificial intelligence and statistics,' ed. W.A.Gale, Addison-Wesley: Reading, Mass. (1986).

4. W.A.Gale and D.Pregibon, An expert system for regression analysis. Proceedings of the 14th Symposium on the Interface, ed. Heiner, Sacher, and Wilkinson, Springer-Verlag: New York, 110-117 (1982).

5. S.M.Gore, I.G.Jones, and E.C.Rytter, Misuse of statistical methods: critical assessment of articles in BMJ from January to March 1976, British Medical Journal i, 85-87 (1977).

6. L.Hakong and F.R.Hickman, Expert system techniques: an application in statistics. In 'Expert systems 85', ed. M.Merry, Cambridge University Press: Cambridge, 43-63 (1985).

7. D.J.Hand, Statistical expert systems: necessary attributes, Journal of Applied Statistics, 12, 19-27 (1985).

8. D.J.Hand, The role of statistics in psychiatry, Psychological Medicine, 15, 471-476 (1985).

9. D.J.Hand, Choice of statistical techniques, Bulletin of the International Statistical Institute, Proceedings of the 45th Session, Vol.3, Amsterdam, 21.1-1 to 21.1-16 (1985).

10. D.J.Hand, Artificial intelligence and psychiatry, Cambridge University Press: Cambridge (1985).

11. D.J.Hand, Expert systems in statistics, Knowledge Engineering Review, 1, 2-10 (1986).

12. D.J.Hand, Patterns in statistical strategy, In 'Artificial intelligence and statistics,' ed. W.A.Gale, Addison-Wesley: Reading, Mass. (1986).

13. D.J.Hand and B.S.Everitt (eds.), The statistical consultant in action, Cambridge University Press: Cambridge. (1987).

14. D.J.Hand and C.Taylor, Practical multivariate analysis of variance, Chapman and Hall: London (1987).

15. R.Hooke, Getting people to use statistics properly. The American Statistician, 34, 39-42 (1980).

16. M.J.Norusis and C-M.Wang, The SCSS conversational system, The American Statistician, 34, 247-248 (1980).

17. R.O'Keefe, An expert system for statistics, Paper presented at the Technical Conference on Theory and Practice of Knowledge Based Systems, Brunel University (1982).

18. D.Pregibon, A DIY guide to statistical strategy, In 'Artificial intelligence and statistics,' ed. W.A.Gale, Addison-Wesley: Reading, Mass. (1986).

19. D.Pregibon and W.A.Gale, REX: an expert system for regression analysis, COMPSTAT-84, Prague, Czechoslovakia (1984).

20. A.M.R.Smith, L.S.Lee, and D.J.Hand, Interactive user-friendly interfaces to statistical packages, The Computer Journal, 26, 199-204 (1983).

21. R.A.Thisted, Representing statistical knowledge for expert data analysis systems, In 'Artificial intelligence and statistics,' ed. W.A.Gale, Addison-Wesley: Reading, Mass. (1986).

22. G.B.Wetherill, C.Daffin, and P.Duncombe, A user-friendly survey analysis program. Bulletin of the International Statistical Institute, 45th Session, Vol.3, Amsterdam, 20.4-1 to 20.4-14 (1985).

23. S.J.White, Statistical errors in papers in the British Journal of Psychiatry. British Journal of Psychiatry, 135, 336-342 (1979).

2 Data Analysis as Search

David Lubinsky and Daryl Pregibon
AT&T Bell Laboratories, Statistics and Data Analysis Research Department
Murray Hill, New Jersey

1. Introduction

Data analysis is an art practiced by individuals who are skilled at quantitative reasoning and have much experience in looking at numbers and detecting patterns in data. Usually these individuals have some background in statistics.

Nowadays data analysis is done via interactive computing, ideally in an *environment* well-suited to needs of the analyst which includes facilities for data management as well as numerical, statistical and graphical algorithms and functions. The analyst loosely follows what we call the *display/action* cycle, in which one looks at some display (some numerical or graphical summary) and chooses an action on the basis of their interpretation of that display. This will usually involve some numerical computation or data manipulation ultimately resulting in another display, and so on. Looking over the shoulders of several skilled data analysts reveals the individuality of their art, both in their choice of techniques and the order of their application. As with all art, there is seldom a single best rendition, but good from bad is easily distinguished.

We have been concerned with trying to incorporate data analytic skills into so-called knowledge-based "expert systems". The idea is to capture the *heuristic* knowledge that data analysts employ and code it into software which can be made widely accessible. In this endeavor, we have come to view data analysis as much more structured than the 'over-the-shoulder' view described above. In particular, we believe that much of the artistry of skilled data analysts is concerned with guiding a *search* of the space of *descriptions* that they will use to summarize their efforts. We believe that a more precise and explicit enumeration of this space, together with its *logical* structure, can be usefully exploited in an automated system for data analysis.

This does not mean that we are attempting to mimic how a skilled analyst goes about his business. This approach was used in an earlier prototype system for regression called REX (Gale and Pregibon, 1984). A major limitation of REX and other like systems which attempt to model data analysts is that they are not able to exploit the context which gave rise to the data in the same way as data analysts do. Part of the problem is shared by expert systems in any domain: how to represent and manipulate real world and common sense knowledge in applied problem solving. The other part is specific to expert systems for statistics: which aspects of the context are relevant, where are they relevant, and how do they influence the course of an analysis. These are difficult questions to answer in general and even more difficult to usefully represent in software. This situation is unlikely to change in the near to distant future.

Our current solution is not to try to *include* context, but rather to *accommodate* it by providing a collection of good descriptions, and letting the user/analyst choose the particular description that is consistent with the context. This may be inefficient but not an entirely new idea in data analysis. In particular, the problem of *variable selection* in multiple regression provides a leading case where context-free searching for good descriptions (*i.e.* regression equations) is done, providing the analyst with a collection of descriptions of varying complexity and accuracy. The analyst can then browse the collection and choose as seems fit given the context and other auxiliary information such as the cost of collecting each variable and the apparent accuracy/reliability of each. We believe that this method is generally useful and seemingly necessary in any automated data analysis system.

Thus our current software attempts to attain the same *end* as the (human) analyst, but using quite different *means*. In some ways we believe that by exploiting computers for what they are currently good at, *i.e.* search, rather than by trying to make them emulate what humans are good at, *i.e.* thinking, that such systems may actually perform better than some human analysts. Indeed, this phenomena has been observed in the domain of computer chess where the best programs currently do not mimic the masters, but rely on computational power and specialized hardware to look ahead to many times the number of board configurations than a human could possibly do. To date, no computer has yet beaten a grand master, but they can beat many of us mortals. We think the same could be true of data analysis.

This paper is organized as follows. Section 2 provides a brief overview of some early ideas on doing massive computation for regression analysis. Section 3 introduces an example to motivate the range of descriptions that statisticians commonly use to describe regression data. In Section 4 we formalize our thesis of "data analysis as search" and outline the three requisite components of search as they relate to data analysis. In Section 5 we describe our new prototype system called TESS (Tree-based Expert Systems for Statistics). In Section 6 we further distinguish between our approach and that of conventional expert systems for those readers unfamiliar with the latter. The paper is concluded with an Appendix which contains a critique of our earlier attempts at building statistical expert systems and a supporting vote for our current approach.

2. The Grand Masters of Data Analysis

The textbook *Fitting Equations to Data* by Daniel and Wood is unrivaled in their treatment of regression analysis. It is unique in many ways:
- they use case studies, not to motivate individual *techniques*, but rather to motivate how these techniques are combined to form an analysis *strategy*
- they introduce a flow diagram to describe their strategy *explicitly* (on the page facing page 1!)
- they use simplicity as well as goodness of fit to choose good equations
- they use graphical and numerical techniques in a synergistic fashion to sort out complicated structure
- they *systematically* explore the joint effect of individual characteristics of the data.

We first thought that their key contribution to our work was their diagrammatic representation of their analysis strategy. It is indeed important, not so much for details of their approach, but rather for the fact that they make it explicit. On further reflection however, we felt that the key ingredient was their insight that (Daniel and Wood, Section 5.7)

> 'Although we usually get our notions one at a time, it does not follow that we
> will find the best equation(s) by accepting or rejecting each notion after exa-
> mining it *once*. It will often be possible to broaden our experience by looking
> at each new idea in the full context of all its predecessors.'

As pertains to fitting equations to data, they mean that a sequential, one-pass (linear) search for a good fitting equation is not the best we can and should do. And yet this is what most data analysts currently do since we can only reasonably carry out a *single* analysis at a time. We thus commit ourselves to certain decisions which are seldom challenged later on when additional features of the data become apparent. Thus we might decide straightaway to remove an apparent gross outlier, and subsequently add a squared term in x, and allow for heterogeneous variances by including weights, when taking logarithms of y and doing ordinary (unweighted) least squares on x accomodates the outlier, nonlinearity and heterogeneity in a single blow.

Daniel and Wood use the above argument as motivation for doing an extensive search for good equations that even today is quite remarkable. They achieve their end by using computing power where human capabilities become overwhelmed. Specifically, they implemented their suggestion as follows:

● they identify the features that they choose to describe
{linearity of regression, the subset of regressors, and the presence of outliers}

● they determine which features are individually exhibited in the data
{using diagnostic techniques and component-plus-residual plots}

● for these k features, they devise a 'transformation' which should eliminate this feature
{data reexpressions, removing variables, and removing observations}

● they fit all 2^k equations (or some fraction thereof)

● they compare fits using a variety of analytical and graphical tools
{ANOVA for comparing residual sums-of-squares and residual plots for assessing homogeneity, normality, etc.}

Daniel and Wood's implementation shifted some of the burden of doing a comprehensive search for good descriptions from the analyst to the computer. They downloaded to the computer those things (massive computation) that it was good at, and kept to themselves those things which computers were not good at (pattern recognition). As such their implementation was still very analyst intensive.

The weak link in their implementation was their inability to include, up front, all the features of a regression equation which were ultimately of interest to them. Influence points, normality of residuals, variance homogeneity, and residual correlation ('nesting' in their terms), were all handled manually. We conjecture that the main reason they decided to treat these features outside the computerized search procedure was because their measure of goodness, C_p (or RSS_p), was not able to deal with them.

In the time since publication of their book, flexible interactive software environments have replaced batch processing, computer memory is no longer a serious limitation, and computer speed has increased dramatically. Yet data analysts still carry on with their linear search for good descriptions. Clearly the analyst's capability to manage *multiple* (parallel) analyses is the limiting factor. Even multi-window terminals have not helped to fundamentally change our mode of analysis. We argue that an update to Daniel and Wood's implementation can effect the kind of fundamental change we feel is necessary.

In the next section we abstract from their implementation what we feel are the basic ingredients that generalize their insight to other data analytic tasks.

3. The Basic Ingredients of Data Description

Just what constitutes a description? Consider the data of Koenker and Bassett (1980) on the relationship between sales (y) and size of sales force (x) for 108 advertising firms (see Figure 1). Some descriptions of these data are:

☞ the data are given by the 108 ordered pairs (3636,2642), (810,1002),..., (101,90)

. . .

☞ there are two groups of points, one in which the relationship of sales to size is linear, sales = 9.86 + 0.68 size, with residual standard error 66.7, and the other consisting of the single point (2810,1002).

☞ the relationship of sales to size is linear, sales = 22.03 + 0.64 size, with residual standard error 107.

☞ the relationship between sales and size is multiplicative, sales = 1.20 sales **.911, with relative residual standard error of 0.105.

☞ there are two groups of points, one in which the relationship of sales to size is multiplicative, sales = 1.13 sales **.923, with relative residual standard error of 0.103, and the other consisting of the single point (2810,1002).

. . .

☞ there is no relationship between sales and size; the marginal distribution of size is highly skewed to the right with median 246.5 and quartiles 141.5 and 583.5; the distribution of sales is also highly skewed to the right with median 175 and quartiles 112 and 410.

These six descriptions are graphically displayed in Figures 1-6. They illustrate a number of characteristics of how we describe data.

1) The elements of a description are both qualitative and quantitative in nature. Typically the qualitative components refer to descriptive *features* of the data, and the quantitative terms refer to descriptive *statistics* implied by the features. Thus <linearity> is a feature, and <intercept, slope> is a statistic suitable to describe this feature.

2) Redundant features, due to logical implications, are suppressed in a description. For example, since <normal> implies <symmetric and unimodal>, we don't include the latter features in a description of a normal-looking residuals.

3) Data reexpression is commonly used in order to make otherwise complicated descriptions simpler. The number of such reexpressions used in practice is quite small.

4) Data splitting is commonly used in order to make otherwise complicated descriptions simpler. The number of splits of a data set is usually quite small.

5) In practice nothing is ever <exactly such and such>, so we implicitly mean <approximately such and such> whenever we say <such and such>. In the event that a data set is exactly symmetric say, it may or may not be important to explicitly include the exactness in the description.

There are (at least) two important dimensions upon which different descriptions can be compared; we call these *accuracy* and *interestingness*. Accuracy pertains to the agreement between the data and its description. Interestingness pertains to both the verbosity of the description and its usefulness in conveying information. Thus, in the hypothetical listing of descriptions above, the first one, complete enumeration, is the most accurate and the least interesting. The middle four are all less accurate but certainly more interesting. The final one is interesting but woefully inaccurate.

Most of the above considerations are extensions of ideas in Daniel and Wood. We attempt to formalize these in the next section. In doing so we hope to exploit the explicit structure for subsequent computer implementation.

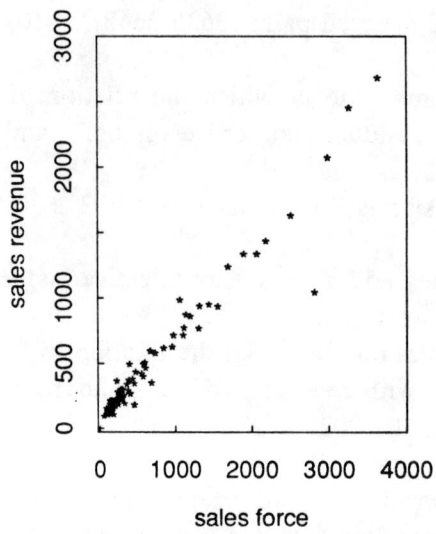

Figure 1.

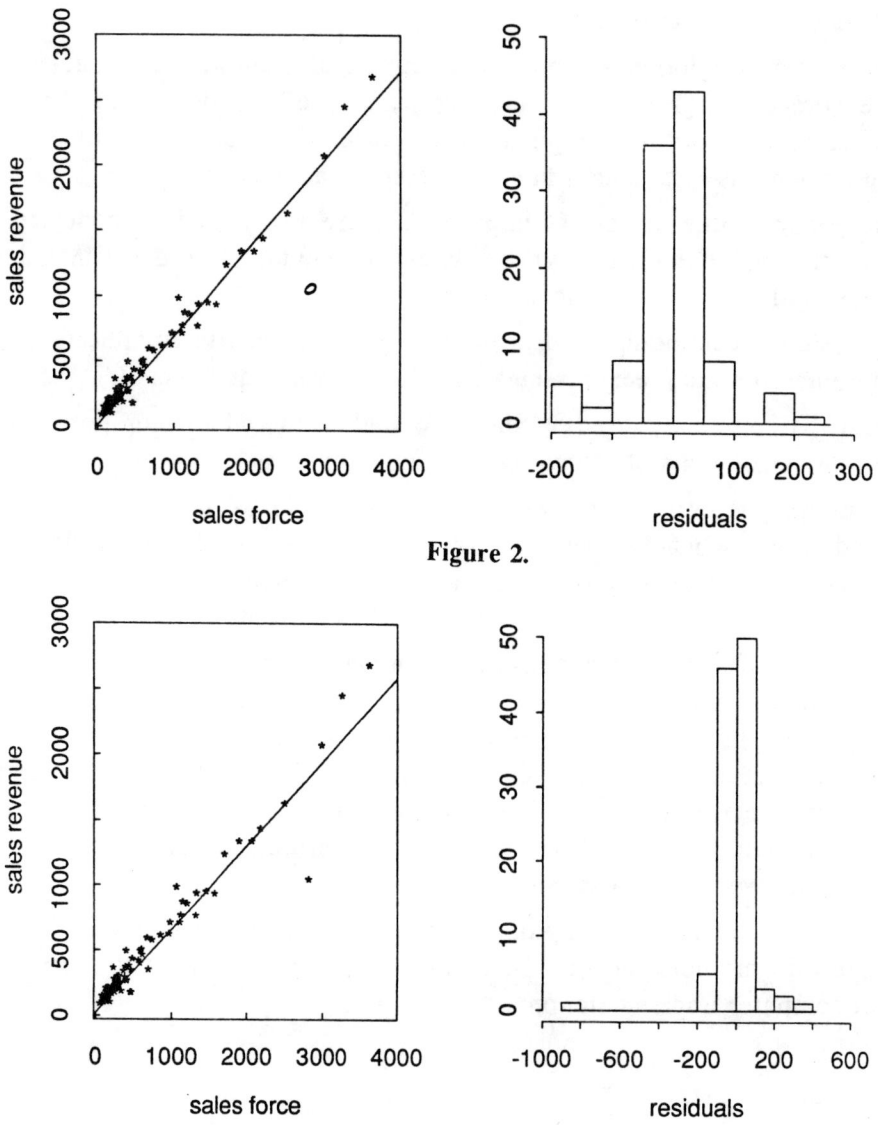

Figure 2.

Figure 3.

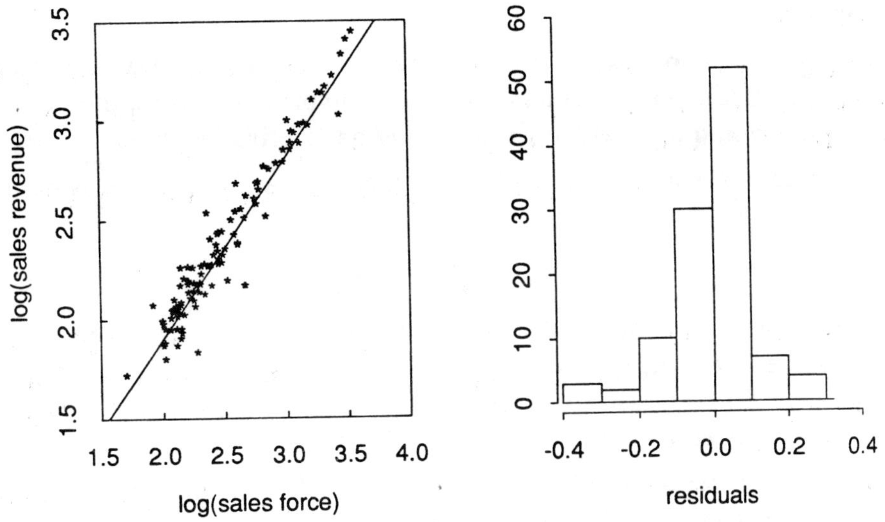

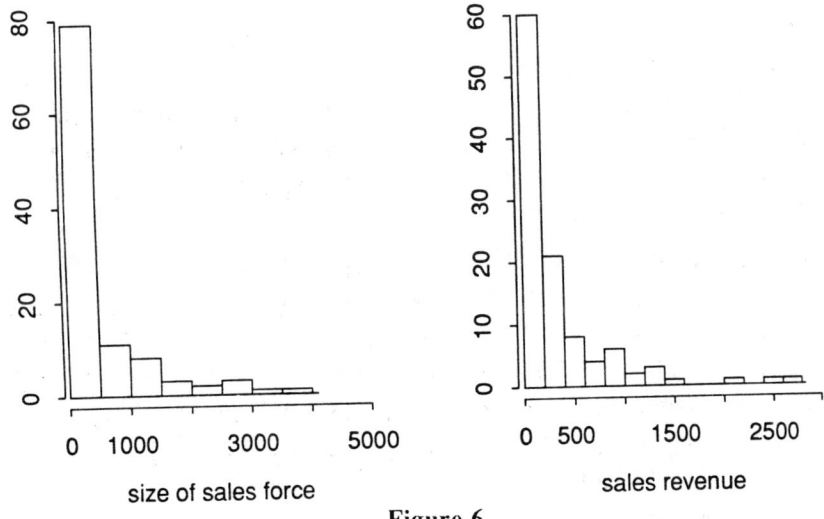

Figure 4.

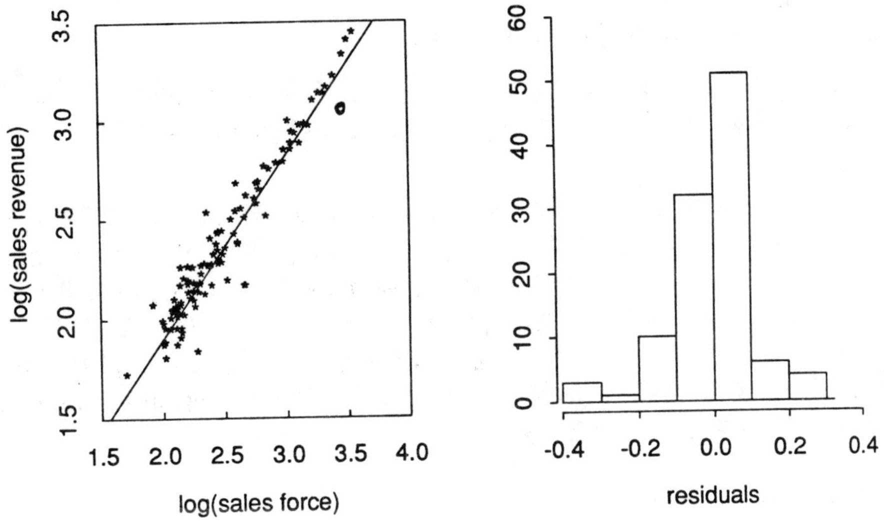

Figure 5.

Figure 6.

4. Formal Development

Our treatment here is similar to that of Mallows (1983). Whereas he was concerned with developing a probability-free theory for quantitative descriptors, *i.e.* descriptive statistics, we are primarily concerned with qualitative descriptors, *i.e.* descriptive features.

Consider a space of data sets Y and a space of descriptions D. A descriptor δ is defined as a mapping

$$\delta : Y \to D$$

such that $\delta(y) = d$ is a description of the data set $\mathbf{y}$. Data analysis is concerned with searching the description space D for mappings which provide good descriptions of the data set $\mathbf{y}$. By 'good' we mean that the the description captures the salient features of the data *and* that the description is not unduly complicated.

Data analysts often perform *transformations* on data, say $\mathbf{y} \to \mathbf{y}^*$, and then proceed to describe $\mathbf{y}^*$. Often this results in great simplification of description. The two common transformations used in data analysis are

- data reexpression, $\rho : Y \to Y$, 1-1, onto
- data splitting, $\sigma : Y \to \{Y_1, Y_2\}$, disjoint subspaces.

We can include descriptions of transformed data in the above framework by defining $\delta_\rho(\mathbf{y}) = \delta(\rho(\mathbf{y}))$, since to describe a 1-1 function of $\mathbf{y}$ is equivalent to describing $\mathbf{y}$ itself, and $\delta_\sigma(\mathbf{y}) = \delta(\mathbf{y}_1) \bigcup \delta(\mathbf{y}_2)$.

In order to automate this view of data description, it is necessary to enumerate the search space D, attach a 'goodness' function γ to each member of D, and define an appropriate search strategy. Factorization of the problem into these distinct components can also prove useful in capturing the different ways people think about data analysis. In particular, we think that the description space and search strategy for a specific task should be fairly constant for all data analysts, but that the 'goodness' function will vary depending upon particular data analysis goals and analyst preference.

4.1. Enumerating the Search Space

In order to enumerate the search space it is necessary to structure the space of descriptions D. The descriptive features of data sets in Y and allowable transformations are the building blocks of D.

Because of the logical implications that relate features and because transformations are applied at specific points in an analysis, we find that an (inverted) *tree* provides a concise way to represent D. A description is a *path* through the tree, starting at the root, but not necessarily ending at a leaf. Each node in the tree represents either a descriptive feature or a data transformation. An example of such a tree is displayed in Figure 9 which will be described in detail below.

The key feature of the tree is its hierarchical structure. Nodes lower in the tree not only imply the features of their ancestors, but they are at the same time, less accurate and more interesting descriptions. The nodes with semicircular sides correspond to transformations, which operationally correspond to attaching *subtrees* beneath them. Thus the apparent shallowness of the tree is misleading as it is in fact infinitely deep (allowing for repeated transformations *ad infinitum*), though in practice, we have yet to find an example where an infinitely complex description is useful.

The strategy for a particular data analysis task may be expressible in terms of a single tree, as in the case of univariate batches shown in Figure 9. For more complex tasks it is usually

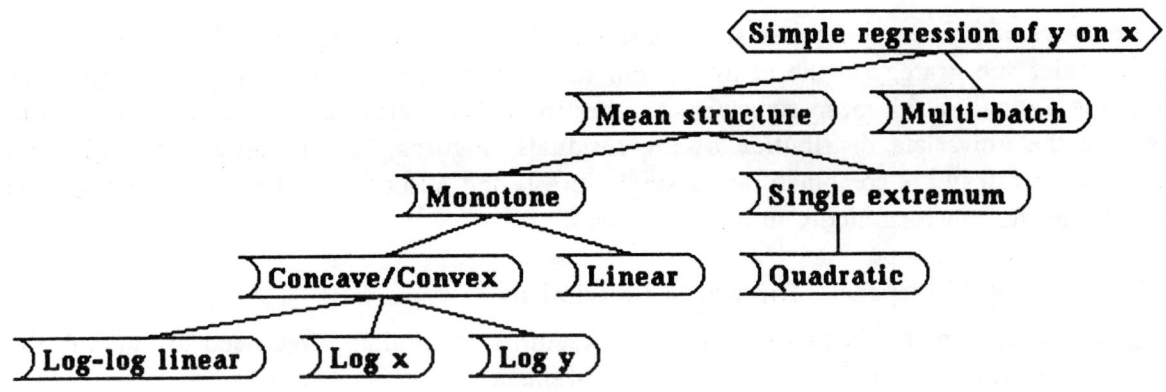

Figure 7.

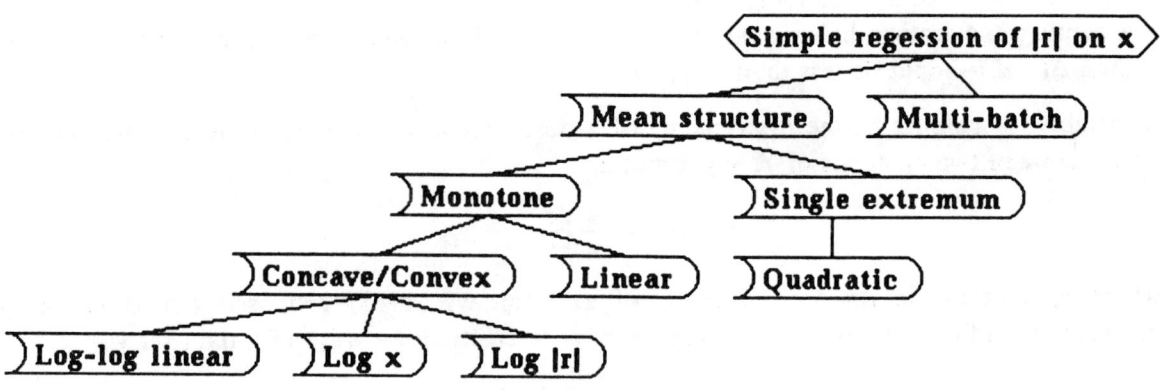

Figure 8.

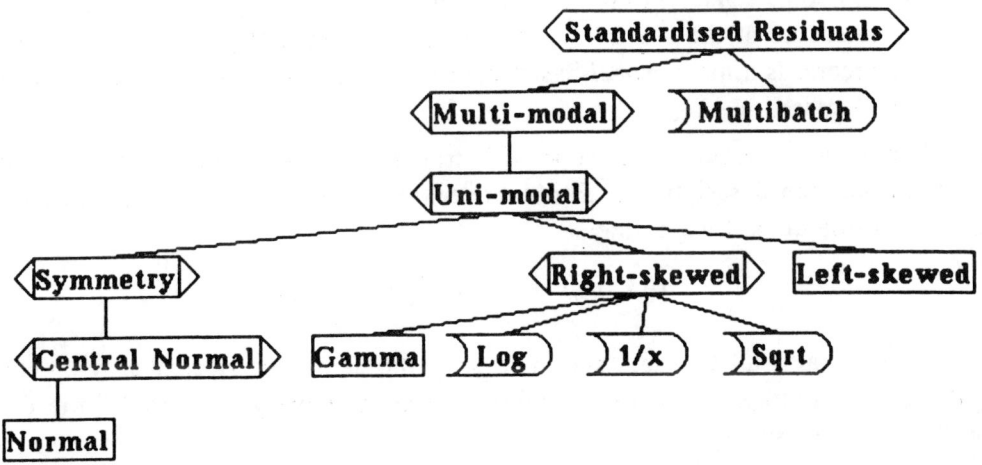

Figure 9.

more convenient to express the strategy in terms of a few interacting trees. Each tree handles a particular sub-strategy of the analysis. For example, in regression analysis, there could be one tree to analyze the mean structure, another to analyze variance structure, and a third to analyze the univariate distribution of the residuals (Figures 7-9). In these cases, all trees used by a strategy are combined into a single forest, and all nodes in the forest are available as subtrees to transform nodes in the same forest.

4.2. Attaching Discrepancy Measures to Descriptions

A good description should capture the salient features of the data. We need to quantify the degree to which a description satisfies this requirement.

A discrepancy measure is some function $\Delta : Y \times D \rightarrow R$ which maps the discrepancy of a description $d(\mathbf{y})$ for the data set $\mathbf{y}$ into some real number. In order for it to be useful as a means to guide search of D, it must

- be comparable across features and transformations (*i.e.*, nodes of the tree) which are the building blocks of descriptions, and
- capture the feature being quantified so that if f_i is indeed a feature of $\mathbf{y}$ and f_j is not, then $\Delta(\mathbf{y}, d_i)$ should be less than $\Delta(\mathbf{y}, d_j)$.

Invariably, a measure of the discrepancy of a description is based on some measure of distance between two elements of Y, say $\mathbf{x}$ and $\mathbf{z}$,

$$\pi(\mathbf{x}, \mathbf{z}) = \frac{1}{n} \sum (x_i - z_i)^2$$

where x_i denotes the ith order statistic of $\mathbf{x}$. Mallows (1983) proposed two discrepancy measures based on π to assess the accuracy of a description $d = d(\mathbf{y})$ of a data set $\mathbf{y}$:

$$\Delta(\mathbf{y}, d) = \max_{\mathbf{x} \in Y_d, \mathbf{z} \in Y_d} \pi(\mathbf{x}, \mathbf{z})$$

where Y_d is the inverse image of the descriptor δ, *i.e.*, $Y_d = \delta^{-1}(d(\mathbf{y}))$ contains all the data sets in Y with the same description as $\mathbf{y}$, and

$$\Delta_0(\mathbf{y}, d) = \min_{\mathbf{z} \in Y_d} \pi(\mathbf{y}, \mathbf{z})$$

where $\mathbf{z}$ is in some sense, an *estimate* of $\mathbf{y}$, or some most typical element of Y_d. The first of these is a measure of the *size* of Y_d, corresponding essentially to its maximal squared diameter. The second is a more *model*-based measure, somewhat akin to our usual notion of residual mean square.

Any discrepancy measure must cope with transformations in order to allow comparisons to be made between descriptions. For data reexpressions, all competing descriptions of $\mathbf{y}$ will be in the same units if we define

$$\Delta(\mathbf{y}, d_\rho) = \max_{\mathbf{x} \in Y_d^*, \mathbf{z} \in Y_d^*} \pi(\rho^{-1}(\mathbf{x}), \rho^{-1}(\mathbf{z}))$$

where Y_d^* contains all the data sets in Y with the same description as $\mathbf{y}^* = \rho(\mathbf{y})$. For data splitting, the discrepancy of the (whole) data set is merely the sum of the discrepancies of the parts, say $\mathbf{y}_1$ and $\mathbf{y}_2$,

$$\Delta(\mathbf{y}, d_\sigma) = \Delta(\mathbf{y}_1, d_1) + \Delta(\mathbf{y}_2, d_2).$$

4.3. Attaching Verbosity Measures to Descriptions

A description can be written as $d = <f, s_f>$ where f refers to descriptive features and s_f to descriptive statistics implied by f. Descriptions are interesting when they employ commonly used terms to convey meaningful features of the data and when they are not unnecessarily detailed. We use the term *verbosity* to capture the amount of detail in both f and s_f.

It seems reasonable to require the total verbosity v to be a sum of v_f and v_s, but that the components could be defined differently. For example, we could have $v_f = \text{depth}(f)$ and $v_s = \text{length}(s)$. This tree-based definition of feature verbosity seems to be acceptable for transformations and non-redundant features, but misses the mark when parent features are implied by their children. The simpler, $v_f = \text{length}(f)$, would seem to handle the problem, in which case we can write $v(d) = \text{length}(d)$.

Other measures of verbosity can be based on information theoretic principles, where rather than counting the number of elements of a description d, one counts the number of *bits* required to store it. We plan to explore this approach in more detail.

4.4. Trading-Off Discrepancy and Verbosity

The data themselves are the most accurate description of the data, and also the most verbose. Similarly, single number summaries are the least accurate and least verbose descriptions. Clearly our notion of goodness of description requires some trade-off between these two properties which makes for interesting and non-trivial descriptors. This is directly analogous to the trade-off between variance and bias for estimators, where mean squared error provides a single summary combining the two.

Currently we have no theory to combine discrepancy and verbosity into a single measure of goodness of description. We suggest that both be computed and the resulting values displayed in a plot of $\Delta(\mathbf{y}, d)$ versus $v(d)$. In this fashion a variety of descriptions of varying discrepancy and verbosity can be compared for the same data set.

4.5. On Search Strategies

Our current approach to regression is not to provide *the* regression of y on x. We require *multiple* answers (descriptions), most importantly because we are doing context-free search and cannot rely on an algorithmic procedure to choose the single best regression description across all contexts.

For a single tree many optional search strategies are possible including breadth-first and depth-first exhaustive search. The regression domain provides an example where exhaustive search is not feasible and some alternative must be used. One solution is to use some *heuristic* search technique that exploits the topology of individual trees and their interaction with other trees in the forest.

The trees represent the space of descriptions for regression data so that with increasing depth comes less accurate but more interesting descriptions. For a particular data set, actual accuracy values are realized as descriptions are generated from D. Realized descriptions can then be represented as points in the accuracy and interestingness plane. Ideally our search strategy should explore the frontier of this two dimensional space since we want accurate, interesting descriptions.

Descriptions on the frontier can arise from minute differences in detail near the leaf nodes of the forest, or they can arise from gross differences near the root node. In regression, where interest centers primarily on mean structure, secondarily on variance structure, and lastly on the distribution of residuals, it is natural to require the search strategy to reflect these priorities.

A breadth-first search of the description space would allow widely different descriptions of the mean structure to be explored before settling down on second and lower order effects in the data. But we do want answers and a breadth-first search of D is not guaranteed to produce any answers in a finite amount of time. On the other hand a depth-first search is guaranteed to give answers but these would differ in detail at the least interesting aspect of the problem, namely details of shape of the residual distribution.

Our current approach is a modified depth-first search whereby successive descriptions are not enumerated at the leaves of the residual tree but rather at the leaves of its predecessor trees. The idea is that we want to guarantee competing (complete) descriptions in a fixed amount of time. The heuristic search procedure can be regarded as a scheduler whereby regressions are waiting to be run and those that are most likely to be on the frontier get first crack at the resources. The standard 'regression of y on x' is then replaced by '5 minutes of regressions of y on x' whose output can be displayed in the accuracy and interestingness plane.

5. Implementation

We have been actively involved in implementing the above ideas into a Tree-based Expert System for Statistics (TESS). TESS is designed to be used by two classes of people; expert statisticians, who design strategies, and nonstatisticians, who apply the encoded knowledge to describe data sets. We will discuss here only those features designed to be used by the statistician during the implementation phase, that is, while defining the strategy.

5.1. Subjective Mappings

Each node on a tree corresponds to a particular feature that could be present in a data set. When defining the tree, the statistician must define a test for each node. This test must capture the concept represented by the node. That is, for any two data sets if the value of the test is greater for the first than the second, then the associated feature should be more strongly present in the first. If these test statistics are all defined on a single uniform basis, such as likelihood, then the values will be directly comparable across nodes. It may be difficult however to calculate likelihood statistics for each feature of interest. For example, while it is possible to define a likelihood statistic for symmetry, it is neither easy to derive or compute. The approach taken in TESS is to allow the statistician to define arbitrary statistics for each feature and then to subjectively transform these statistics to a common basis. We will discuss this process in greater detail below, but first we describe how a statistician would set about implementing a strategy.

5.2. Implementing a Strategy

The process of defining a data analysis strategy passes through the following steps:
  ● Clearly define area of interest (TESS strategies can easily be specialized, but generalization is much more difficult)
  ● Define sub-strategies of the main strategy where appropriate
  ● For each sub-strategy, design a tree which matches as closely as possible to intuition of how data sets should be analyzed
  ● For each node in the resulting trees fill out the node slots (see below)
  ● Choose a collection of typical data sets for this analysis type
  ● Calibrate tests to give reasonable descriptions of these data sets
  ● Refine tree by moving or adding nodes if necessary, and deleting redundant nodes
  ● Finally test strategy on a new set of data sets

It is clear that many of these steps are far from trivial, and that defining a strategy involves a major investment for a statistician. In this section we discuss in more detail the range of information that must be encoded in each node and the tools that TESS provides in order to simplify the development.

5.3. Node Slots

Recall that there are there are two types of nodes: "Test Nodes", and "Transform Nodes". The test nodes are drawn with angular sides, while the transform nodes have semi-circular sides. The symbols used in drawing the nodes are mnemonic in the sense that test nodes are similar to the test symbol originally used in flowcharting, while the sides of the transform node denote pointing or sending of the data to another node.

Each node has a number of slots associated with it. The "test" slot defines a statistical test which returns a value between 0 and 1, depending on the degree to which the characteristic which it tests is present. The "plot" slot points to a procedure which displays visual information specific to the characteristic being tested at the node. The "name" slot is a single word description of this characteristic. Corresponding to f and s_f, introduced in Section 4.3, each node has a "description" slot consisting of parameterized text fragments. These are used in building up an English description of the data set as it passes through the tree. So for example, a node named "Normality" could use the correlation between the order statistic of the data and normal quantiles as a test of normality, a normal probability plot (or some deviant thereof) for display purposes, and the text "the *data* are normal with mean *mean* and *variance*" for verbalizing the description.

In addition to the above, transform nodes have other important slots. The "transformation" slot is an expression which calculates a new data set from a given data set. This new data set is then sent for further analysis to the node named in the "address" slot.

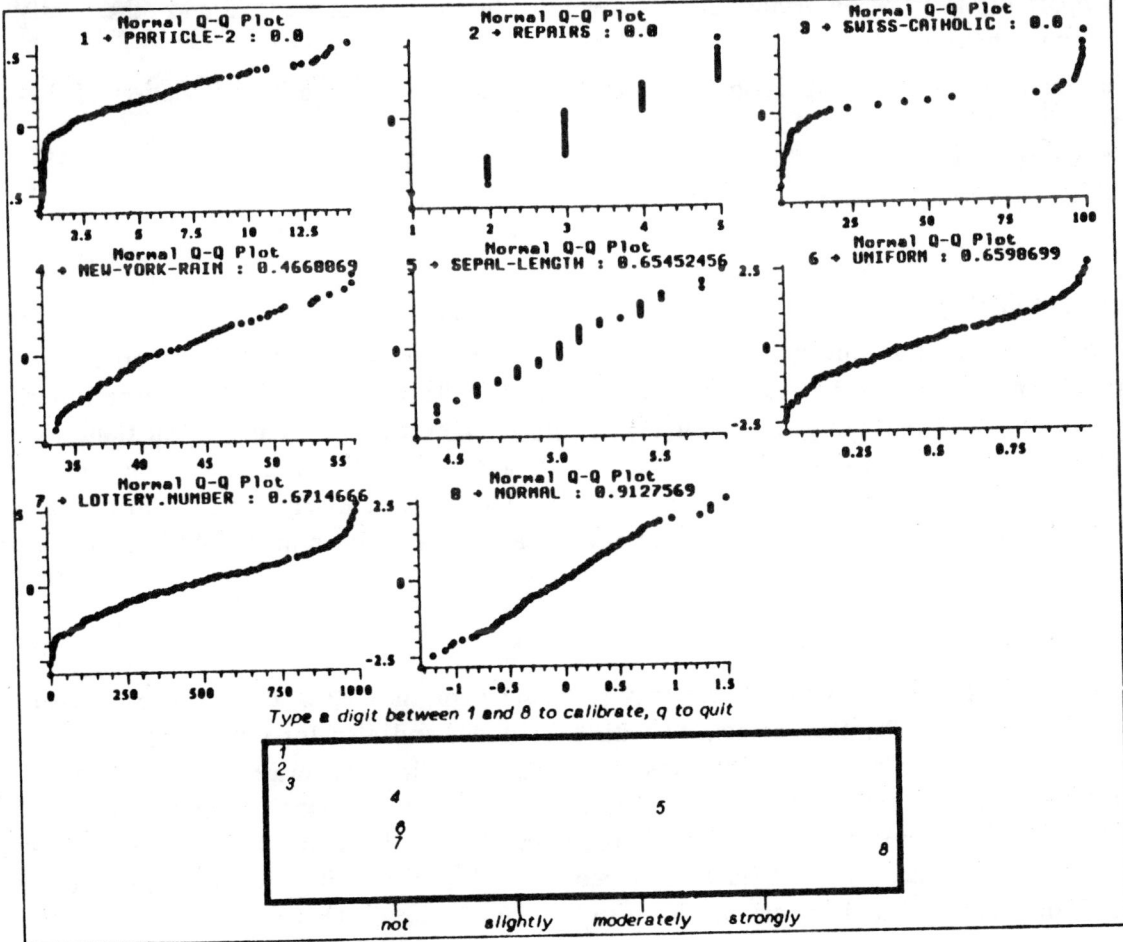

Figure 11.

5.4. The Environment

The screen dump in Figure 10 shows TESS in the basic mode of interaction. There are four windows on the screen. The largest window at the top always displays the current strategy tree. The narrow window in the middle left shows all the commands that are known to the system, while the window on the lower left shows the slots associated with the current node, that highlighted with inverse video in the tree window. The final window on the lower right is a utility area, used for displaying plots. Since it is also an active Lisp window it can be used for scratch calculations in Lisp.

5.4.1. Tree Editing

It should be apparent that the trees defining the strategy are central to TESS, and also that as the statistician goes through the process of defining the tree, many changes will be made before the tree is a satisfactory representation of the statisticians mental model. To facilitate these changes, TESS provides a complete set of interactive tree editing tools which allows the statistician, using the mouse, to edit the tree.

5.4.2. Vector Language

TESS is implemented on a Lisp machine, a rather hostile environment to statisticians, since its primary language is Lisp. Lisp has many strengths for symbolic computations and data structure manipulations, but for the numeric calculations that comprise statistical algorithms, it is just too unwieldy. Since there are currently no large statistical libraries in Lisp, the statistician using TESS must program all of the tests for his strategy. To make this task somewhat more palatable, TESS provides a mini language for statistical computations. This language is based on Lisp, but allows familiar infix algebraic expressions, and also extends all operators to work on vectors or matrices after the fashion of APL, S and other statistical environments.

For example the following defines a function to calculate the standard deviation of the elements of a vector:

```
(function sd(x)
  sqrt(sum((x-mean(x))^2)/(len(x)-1)))
)
```

5.4.3. Caching

In order to speed up numeric calculations, functions which would be expected to use a nontrivial amount of time in execution, can be specially flagged. The values returned by these functions are stored along with their arguments after execution. When a function is called again on arguments for which its value has already been calculated, the value will not be recalculated, but simply retrieved from storage. This is particularly useful with certain utility operations such as ranking which might be called many times in the analysis of a single data set.

5.5. Calibration

An important part of designing a strategy is the selection of a set of data sets to be used as a calibration set for the tests associated with each node, and also for evaluating the tree as a whole. Such a collection of data sets is associated with each strategy and is stored and loaded along with the strategy. The chief criterion that the statistician should use in selecting these data sets is coverage. That is, the data sets should span all of the features which are relevant to the strategy being developed. To achieve this, it is not enough to use only 'real' data sets. Some data sets will have to be manufactured. This can easily be done using random number generators provided by the language. TESS also provides tools for adding and deleting data sets.

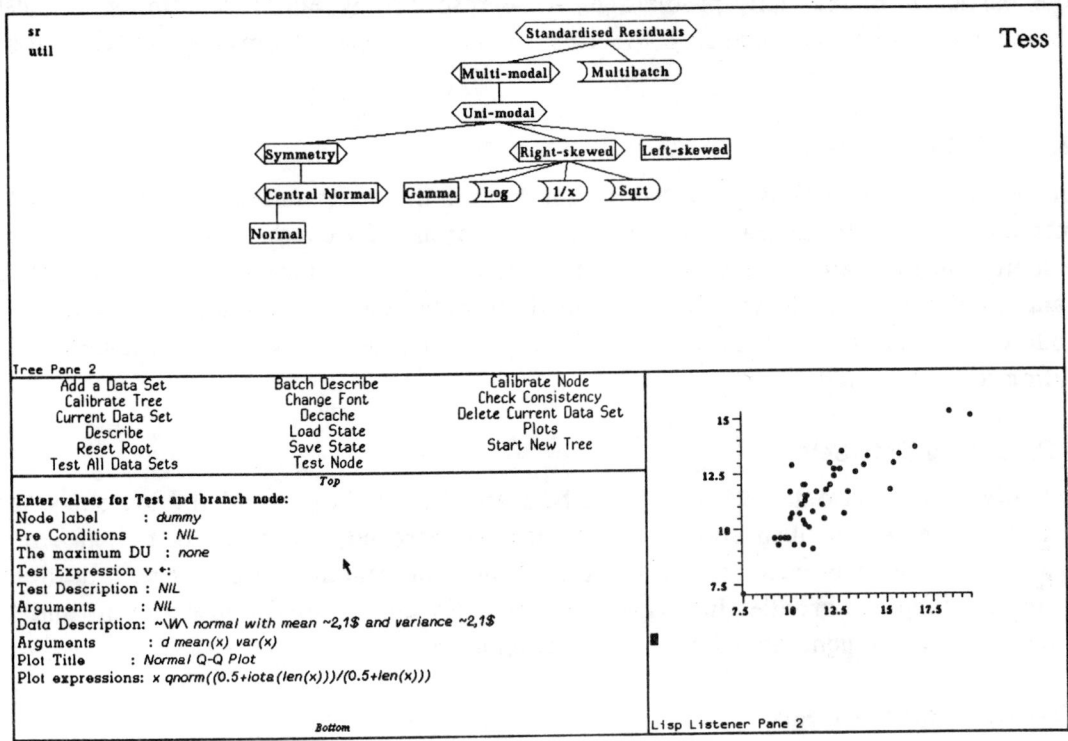

Figure 10.

5.5.1. Local Calibration

The test statistics associated with each node are defined to return values between 0 and 1. In general, we want to allow the statistician to use any statistic that seems appropriate, even if it does not have a well known distribution. This will be especially useful when defining a strategy for an area of data analysis which has not been well studied.

We use the "Normal" node as an example. Let us say that the test associated with this node calculates the correlation between the order statistics of a single batch of data and the appropriate percentiles of the standard Normal distribution. This correlation will always be between 0 and 1, and is as such a valid test. However this correlation will be high for data sets from most distributions, so we need some way to map the correlations onto a scale which is standard for all nodes. To do this, we first define a function which calculates the correlation, then subtract 0.8 and multiplies by 5 to get a value between 0 1 and 1. We then apply this function to a number of data sets and ask the statistician to score how strongly he believes the feature is present in each data set. The responses of the statistician define a mapping from the range of test results into a subjective scale which is common for all nodes. The values on the subjective scale define the accuracy of the description.

Figure 11 illustrates how test calibration is done. The plot for the test is drawn for each selected data set and the statistician scores each in the box at the bottom of the figure. Each plot is numbered and the position of the digit corresponding to the plot indicates how strongly the statistician believes this feature is present in the data set.

If the relationship between the belief values and the test values is not monotone, this implies that the numerical test does not match the concept that the statistician had hoped to capture for the particular node. This problem must be resolved by either redefining the test to match the concept, or restructuring the tree so that the test becomes appropriate for the node.

5.5.2. Global Calibration

A further calibration tool is provided to check that the tree semantics are satisfied, i.e. parents must always be at least as accurate as children. To check that accuracy decreases with depth, we calculate the accuracy (on the subjective scale) for each node for each data set, and then print a table with rows defined by data sets and columns by nodes. Data set/node combinations that violate the consistency requirement must be investigated and the offending tests redefined.

5.6. Describing Data Sets

To actually describe a data set, the "Describe Data Set" is chosen and the system searches through the tree as described above. As the tree is searched, a parallel tree is grown with the test results and generated text fragments. When the search phase is over, the descriptions on the dominant frontier in terms of discrepancy and verbosity are chosen, and complete descriptions are generated from the text fragments.

6. Tess and Expert Systems

Much has been said and written recently about expert systems, and the reader may be wondering where TESS fits into the spectrum of knowledge based systems. In this section we discuss TESS in terms of a number of areas related to expert systems.

6.1. Knowledge Acquisition

One of the major bottlenecks in the production of experts systems has been in acquiring knowledge from the expert in the application domain. Typically two experts are required. One who knows how to use the expert system itself, known as the knowledge engineer, and one with expertise in the area of interest, the domain expert. The knowledge engineer communicates with the domain expert, goes away and implements his understanding of what he has been told. The expert system is then tested, on a set of test cases and the results evaluated. Usually the results are flawed, so the domain expert and knowledge engineer iterate this process until either the results are satisfactory, or returns have diminished to the point where it is no longer viable to continue.

TESS takes a different approach in that the domain expert, the statistician, is able to enter and debug his knowledge himself. There is no need for an intermediary, and the debug cycle is made very short by the interactive nature of the tools. Using TESS does require a fair degree of facility in the use of computers and statistical programming, but certainly not more than most statisticians are capable of.

We have not tested whether this is in fact an improvement on the approach described above, but it does seem to be more direct. By avoiding the need for a knowledge engineer, we simplify the process of knowledge acquisition, and remove the possibility of errors introduced by misinterpretation of the domain expert.

6.2. Knowledge Representation

A second important area in the development of expert systems, has been the method of representing domain knowledge in the system. The original expert systems and many since then store knowledge in terms of 'if .. then .. ' rules. For example a typical MYCIN (Shortcliffe, 1976) rule is:

```
If infection is meningitis and
infection may be bacterial and
patient is severely burned
then there is suggestive evidence that Pseudomonas aeruginosa is
one of the organisms causing the infection.
```

The knowledge in TESS is not as explicitly encoded, but it could be reformulated in terms of rules like these without any loss of information. For example:

```
If symmetry value > 0.6 and
data contain more than 10 points
and data are unimodal
Then test for normality.
```

However in representing knowledge in this way, the structure encoded in the tree is not immediately visible, and it becomes much harder to debug and analyze. A data analysis strategy represented as a tree is arguably more comprehensible than a flat list containing many rules.

6.3. Explanation

One advantage of rule based expert systems is that they allow explanations to be attached to each rule. This is useful if a user wants to query a rule or requires an explanation for a complex sequence of deductions. In TESS there is at present no such capability. The system is a black box: data sets in, descriptions out. This approach is obviously too restrictive and is an area that requires further work.

6.4. Evaluation

Evaluation of expert systems is always a difficult task, requiring comparison of the systems performance to that of human experts. In our case we could set up a double blind test: give descriptions of data sets from TESS and statisticians, to other statisticians, and asking them to rate each description. TESS's descriptions would of course have to be doctored slightly, so that it would not be obvious that they were machine generated. This could be done without changing their substantive content.

6.5. Reflexive Examination

In one aspect TESS is very different to most other expert systems; the usual mode of operation would be for the expert system to advise some course of action. This may or may not be acted upon in the real world. The results of the action if executed, would usually not be fed back to the expert system.

On the other hand the only world that TESS is concerned with is inside the system (the statistical computations and their results are available to TESS). This means that TESS can give multiple solutions (descriptions) and rank them. This ability leads to exciting possibilities in terms of learning. Given some human input, TESS could improve its data analysis capabilities by adjusting itself to match preferred descriptions of the statistician. As inconsistencies between what TESS believes is the best description of a data set, and the statisticians belief are detected, they can be automatically resolved by readjusting the interestingness weights to minimize some measure of inconsistency.

References

1. Daniel, C. and Wood, F.S. (1980). *Fitting Equations to Data, 2nd Edition.* John Wiley & Sons, NY.

2. Gale, W.A. and Pregibon, D. (1984). "REX: An expert system for regression." *Proceedings COMPSTAT84*, Prague, pp 242-248.

3. Koenker, R.W. and Bassett, G.W. (1980). "Robust tests for heteroscedasticity based on regression quantiles," *Econometrica.*

4. Mallows, C.L. (1983). "Data description." In *Scientific Inference, Data Analysis, and Robustness,* Edited by Box, Leonard & Wu. Academic Press, NY, pp 135-151.

5. Shortliffe, E.H. (1976). *Computer Based Medical Consultations: MYCIN.* American Elsevier, New York.

6. Tukey, J.W. (1977). *Exploratory Data Analysis.* Addison-Wesley, Reading, MA.

Appendix A

Why mimicking data analysts by expert systems?
A. Buja
June 19, 1984

The discussion centered around REX concerns largely the question of how a data analyst acts in specific situations, and how this can be captured in terms of strategies for REX. There is great value in automating strategies of data analysts: it forces him/her to lay out subjective rules and make them the object of study. Benefits might be in the ultimate optimization and fool-proofing of strategies, as well as improved and more objective teaching of data analysis.

However, the question which I want to raise is this: in designing an expert system for data analysis, should we actually follow the analyst's actions faithfully, or should we rather design new sets of actions which are appropriate for an automated analyzer? This more fundamental question is motivated by the fact that the strengths of humans and machines are in different areas. Humans have judgment, -- they are the ultimate recourse if it comes to the question of whether an analysis makes sense or not. Humans are also good at pattern analysis through visual perception. This allows a human to single out reasonable sets of actions from the very outset just by looking, e.g., at a suitable scatterplot.

A machine, however, doesn't share these capabilities at this point, but in some cases it can be programmed to obtain similar results by different means, namely through exploitation of crude machine speed. An example: in analyzing $x-y$ data, visual perception can easily detect concave curvature in the scatterplot, which may suggest a log- or sqrt-transform of x; a machine might not be so good at detecting a qualitative feature such as convexity in noisy data, but it doesn't pose a problem to simply try and compare the correlations between y and x, e^x, x^2, $\log x$, $\sqrt{x}$, x^{-1}, $\cdots$ and finally pick the best one.

In other words: a machine can afford blind search among a finite set of possibilities, where a human singles out one or a few reasonable possibilities. The *search-oriented* approach to an automated data analyzer already has some history in statistics: all-subsets-regression has this flavor, and Daniel and Wood state that they work in regression both with y and $\log y$ from the outset of an analysis. The real intelligence in a searching expert system consists then less of deciding what action to take (as many as possible will be tried), but judging what action leads to good results. And here we have a multitude of standards as expressed, e.g., by RSS, R^2, C_p, Akaike, # outliers detected by a robust fit, dependence on few influential points, straightness of qq-plot of residuals, ...

If searching can be employed as a general strategy for finding models, then at some point the search space might get out of hand. Here we would be forced into heuristic searching, which is at the heart of many AI-efforts: the art of cutting down search space. An example of heuristics in case of models and transformations could be the following: running a cheap smoother and evaluating it at 3 points might give a clue to convexity resp. concavity in the data, and thus eliminate a whole set of potential transformations.

However, before we develop heuristic short cuts, we might first determine how large a reasonable set of models and transforms is.

3 AI and Generalized Linear Modelling: An Expert System for GLIM

J. A. Nelder
Imperial College, London

1. INTRODUCTION

Linear models are a major tool in statistical analysis; generalized linear models (GLM) [6] extend the idea and allow a variety of statistical techniques to be seen as special cases of a general class. These include classical regression and factorial models for continuous data, log-linear models for analysing counts, logistic regression models for proportions, and models with gamma errors etc. Briefly the models may be written in the form

$$E(y) = \mu : \eta = g(\mu) : \eta = \Sigma_j \beta_j x_j$$

Here y is a vector of observations with mean μ, and independent errors from an exponential family, η is a linear predictor formed from covariates x_j with parameters β_j and g( ) is the link function connecting η and μ.

The package GLIM [7] supports the easy specification and fitting of GLMs, together with facilities for vector and scalar arithmetic, sorting, tabulation, graphical displays (scatter plots and histograms), and input/output. There is an interactive language supporting macros (named blocks of instructions) with looping and branching, so that the user can program non-standard analysis in the GLIM language.

GLIM contains no expertise, except in so far as it offers a structured way of analysing data [8]; the user must know what he wants to do and how to do it. This paper describes work in progress to provide a knowledge-based front end for GLIM, the whole to be called GLIMPSE (GLIM + Prolog + Statistical Expertise).

GLIMPSE aims to help with the analysis of data of a kind for which GLMs are suitable. We hope that use of GLIMPSE will improve the user's own skills. At present the user is expected to understand the general ideas associated with GLMs, though later we hope to introduce facilities that will allow the novice user to acquire such understanding.

2. GENERAL ASPECTS OF THE FRONT END

GLIMPSE is a statistical expert system, and such systems differ in several respects from, say, medical diagnosis systems. First the level of abstraction is much higher: GLIM is currently used at more than 1000 sites throughout the world, and has been successfully applied to areas that were never envisaged by its originators. It has also been used to develop new statistical techniques. We can be sure than GLIMPSE will be similarly used, and rightly so. Thus the rules in the front-end must be essentially application-free; this is very different from the rules in a medical diagnosis system.

Secondly, whereas the transactions in a medical diagnosis system are between the user (whether doctor or patient) and the rule-base, in GLIMPSE the transactions are three-sided, involving user, the front-end, and GLIM, the back-end. The information obtained from GLIM may at any time change both the state of the front-end and the state of mind of the user [3].

Thirdly, we need to distinguish between diagnosis (more generally, decision) and inference [1]. GLIMPSE helps the user to find parsimonious models for his data; it does not offer a diagnosis or assign probabilities to the models which emerge. It may (and does) use decision rules in the process, but as auxiliary tools.

2.1 The Case For Libertarianism

Many users regard existing (non-expert) statistical systems as magic black boxes, delivering the (inevitably right) analysis. They treat the system as an authority, and the result is much gross misuse. In our view a statistical expert system like GLIMPSE, if it is to be successful, must not be a black but a glass-sided box. In particular, it must let the user exploit his own special knowledge of his data and of the aims of his study, which, as noted above, may be of a kind unenvisaged by the designers. Thus GLIMPSE is being designed to offer advice and not to give commands; this implies structuring the advice so that at any point it is given conditional on the current state of the analysis, which may have been previously partly or wholly user-driven, rather than system-driven. Not all users will like this kind of system, because it will oblige them to think hard.

2.2 Transparency

Transparency means giving the user access to all the tools that the system itself uses. It is a natural consequence of the libertarian approach; the glass-sided box is a tool box, and the user can open the box and use the tools for himself. The aim should be to allow the user to modify the system's strategies, and indeed to develop and test his own strategies. In

this way expertise will develop to the benefit of all. We want GLIMPSE
to develop into an open system in this sense.

3. TOOLS

The basic tools used in the construction of GLIMPSE are

(i)   GLIM 3.77
(ii)   Sigma Prolog and APES
(iii)   Unix

We consider each in turn

3.1   GLIM - The Algorithmic Engine

GLIM 3.77 [7] is written in Fortran 77 (restricted set) and exists as a
compiled program driven by an interpretative language. The front-end uses
it by constructing GLIM statements for the tasks required and invoking
GLIM to execute them and return the output. Much use is made of macros
and of the dump-and-restore facility, whereby the state of GLIM at any
time may be preserved and recovered as required. The front-end has been
designed to make the interface with the algorithmic engine as clean as
possible, so that another back end could be substituted provided that it had
all the necessary algorithmic facilities required.

3.2   Prolog

Prolog (short for programming in logic) is a declarative language much
used in AI work. It uses as a basic structure the Horn clause in the form

conclusion if condition 1 and condition 2 and ....

The number of conditions may be zero, when the clause becomes a fact.
The elements of the clause contain predicates with zero or more arguments,
so that a clause might take the form

_x sorts-to _y if _x is-permutation-of _y and _x ordered

here 'sorts-to', 'is-permutation-of' and 'ordered' are predicates, and _x and
_y are variable arguments.

Sigma Prolog was developed by Logic Programming Associates [5], and
contains several useful features, including process forking and modules.

Sigma Prolog has a front-end in the form of APES (Augumented Prolog
for Expert Systems) [2], which supports higher-level functions like
querying the user for information not found in the data-base, giving
explanations and providing natural-language templates for these.

3.3   The Operating System

GLIMPSE currently runs under Unix, though it makes rather little use of
Unix's many facilities. The critical facility is the one called FORK in

Unix, i.e. the ability of one process to call upon another to execute a job and return control at the end. We use file-handling features to display text when GLIMPSE runs, while as implementors we use the editing and other relevant features of Unix. Provided that the analogue of FORK is available, we see little difficulty in transferring GLIMPSE to run under another operating system, given a Fortran 77 compiler for GLIM, a C compiler for Sigma Prolog, and the ability of Prolog to invoke GLIM dynamically.

4.   THE TASK COMMAND LANGUAGE

To get things done the user employs a high-level command language, the statements of which define <u>tasks</u>. The elements of a task statement are keywords or variables in the form of numbers, identifiers, or strings. For example, to enter a vector from the keyboard we might type

enter vector A (1 2 3 4 5 6)

Where 'enter' and 'vector' are keywords, 'A' is an identifier (the vector's name) and the brackets contain the list of values. Most tasks require to be executed by GLIM, and so are translated into GLIM statements by the front-end and passed to GLIM as input. The output from GLIM may need to be processed by the front-end before presentation to the user.

Tasks are used by the front end to code its statistical expertise, and the requirement of transparency means that all tasks can also be invoked by the user for his own purposes.

Any task given by the user is checked for syntactic and semantic correctness, and if incorrectly specified the user is helped to amend it accordingly. Users will have varying familiarity with the command language (including none when they first use GLIMPSE), so syntactic help is avialable at three levels, which we now describe.

4.1   Reminder Mode

At any point the user may type '?'. GLIMPSE responds with a syntactic description of forms of possible completion for that task. Thus

enter vector?

would elicit the response

< vector name > < bracketed list of values >

This level of help is for the relatively experience user.

4.2   Prompting Mode

The user types '??' and GLIMPSE responds by asking questions to elicit each subsequent item. Selection of a keyword from a set of possibilities is by presentation of a menu. The final task is presented to the user, who may then ask for its execution or do something else.

4.3   Hand-Holding Mode

The user invokes this mode by

help task

GLIMPSE then uses plain-language questions to elicit the task required, and again presents the completed task. This mode allows the complete novice to set up tasks without any prior knowledge of the command language syntax.

4.4   Discovering The Meaning of Tasks

The special task with initial keyword 'meaning' allows the user to get a plain-language explanation of the meaning of a given task.

5.   HELP IN USING GLIMPSE

The process of analysis is split into a set of <u>activities</u> (see Section 6), between which the user can move freely. Each activity has a set of relevant tasks associated with it, and most have, in addition, statistical expertise encoded as Prolog rules. The user can read text about

(i)   The organization of the activity
(ii)   The tasks available for the activity
(iii)   The statistical strategy for the activity

Such text constitutes the background information, a kind of user manual for that activity. It offers passive help. By contrast the two general tasks

help task
help stats

offer active help, the first with setting up tasks for direct use of GLIMPSE, while the second invokes the statistical expertise encoded in the front end.

At a lower level the user is offered help in answering questions posed by the system. If he types '?' in response to a question this asks for information on the possible acceptable answers. The response '??' produces a menu of further options relating to the question. These include (i) 'expand', which asks for background information on why the question is being asked, and (ii) a request to the front end to give its own suggested answer, for example to the question 'does this plot show a linear relation?'.

It remains to be seen to what extent self-documentation will be sufficient for the GLIMPSE user. There is, of course, no reason why printed versions of the various text files should not be made for off-line reading.

6.   ACTIVITIES IN GLIMPSE

The front end splits the process of analysis into 8 activities, each with a two-letter abbreviation. These are

SE	set environment
DI	data input
DD	data definition
DV	data validation
DE	data exploration
MS	model selection
MC	model checking
MP	model prediction

Brief summaries of their scope follow.

6.1   SE - Set Environment

Here the user may read about the scope of the whole front-end and define his own syntax for tasks.

6.2   DI - Data Input

Data may be entered from the keyboard, or from previously prepared files, either in GLIM format or more general format.

6.3   DD - Data Definition

The types and attributes of variates may be defined, e.g. whether continuous, count or proportion, whether ordered or restricted in values, whether response or explanatory, whether associated with other variates etc.

The quality of help given later will depend crucially on the information given here.

6.4   DV - Data Validation

Carries out checks on data, using information from DI and DD, and reports fully on any failures. Data must be validated before next step and are now 'frozen'. Unfreezing is allowed.

6.5   DE - Data Exploration

Supports exploratory techniques for looking at relationships between response and explanatory variables, and for provisional assessment of link and variance functions in subsequent formal fitting of GLMs. Before exit user must define a data matrix and GLM specification for use by activity MS.

6.6   M S - Model Selection

Finds parsimonious models for linear predictor from data matrix and GLM defined in DE. Produces tree in which extreme nodes correspond to such models. For more detail see Section 7.

6.7   MC - Model Checking

From a model found at MS, applies checkes on internal consistency of fit, including those on scale of covariate, link function, and variance function. Also looks for outliers, influential points, and important terms omitted in original fit.

6.8   MP - Model Prediction

Prediction is here used as in [4]. Produces summary quantities for a model, with measures of uncertainty, for use in subsequent reporting of the analysis.

6.9   Movement Between Activities

The above linear description of the activities does not imply a linear progression by the user through them. Looping is particularly likely between MS and MC. Repeated analysis of the same data set, may lead to short-circuiting of, say, DV in second and subsequent analyses. Some constraints are imposed, however, by requirements that, e.g., data must be validated in DV before being explored in DE. The statistical help will encourage the user not to think of analysis as a linear process.

7.   AN EXAMPLE OF STATISTICAL HELP

We outline the statistical help available at activity MS - model selection.

The input required for this activity has two parts, the required data set and a GLM node. The required data set is a data matrix with a defined response variate and a set of explanatory variates; the user may declare a subset of units to be used in the fitting. The GLM node defines the link and variance functions to be used, also the prior weight and offset vectors if relevant.

The user is first queried about terms needed a priori in the linear predictor. Other explanatory variates are then treated as <u>free terms</u>, whose status (whether or not needed in the model) is to be determined. The terms needed a priori form the initial <u>kernel</u> of the model. A value for the dispersion parameter (σ^2 in normal models) is found, either from the user's prior knowledge or from fitting the maximal model with all terms included. The free terms are now classified by finding for each a forward F-value for adding the term to the kernel and a backward one for dropping it from the maximal model. Each F-value is assessed as large (+) or small (-) and action on the free terms is taken as follows:

Forward F	Backward F	Action
+	+	add to kernel
+	-	leave free
-	+	leave free
-	-	discard from model

A new mode is formed with the revised kernel and remaining set of free terms, and the process is repeated until either the free term set becomes nul or remains unchanged. If nul, then a unique model has been found. If unchanged, then the process branches; each remaining free-term is included in turn and the original process repeated again. This may lead to further branching.

The class of models may be extended by further including compound terms (interactions, quadratic terms etc) and repeating the selection process. It may also be necessary to retrace steps if it turns out that the first estimate of the residual dispersion was considerably in error. The exact rules for this remain to be worked out. Note that the process may not deliver a unique (the best) model; this is highly desirable when several models of almost equal worth occur.

8. TOWARDS AN OPEN SYSTEM

GLIMPSE is not an adaptive system, of the kind where interaction with the users changes the rule set. Users of GLIMPSE are doing their analysis our way. However, the libertarian philosophy offers possibilities for flexible development. Suppose that a user were able not only to give sequences of tasks (as now), but to store such a sequence with a name and a parameter-passing mechanism. The user could then build a new strategy and apply it to various data sets. This new strategy can then be compared with the system's own. If found consistently superior it might eventually replace the original strategy. In such a system new strategies would be limited by the tasks available, unless the user were able to define both the syntax and semantics of new tasks. Such a further extension is in principle implementable.

The speed of development of statistics is high and new techniques are continually arising. Open systems offer the means whereby expert systems can keep up.

REFERENCES

1.   Fisher, R.A. (1956). Statistical Methods and Scientific Inference. Oliver and Boyd, Edinburgh.

2.   Hammond, P. and Sergot, M.J. (1985). apes: Augmented PROLOG for Expert Systems. Logic Based Systems Ltd., 40 Beaumont Avenue., Richmond, Surrey, TW9 2HE, U.K.

3.   Hand, D.J. (1986). Expert Systems in Statistics. The Knowledge Engineering Review, 1, 2-10.

4.   Lane, P.W. and Nelder, J.A. (1982). Analysis of Covariance
      and Standardization as Instances of Prediction. Biometrics, 38,
      613-621.

5.   McCabe, F.G. et al. (1984). Sigma-Prolog 1.0: Programmer's
      Reference Manual. Logic Programming Associates Ltd., 10
      Burntwood Close, London, SW18 3SU, U.K.

6.   McCullagh, P. and Nelder, J.A. (1983). Generalized Linear
      Models. Chapman and Hall, London.

7.   Payne, C.D. (Ed). The GLIM System: Release 3.77.
      Numerical Algorithms Group, Oxford, England.

8.   Thisted, R.A. (1986). Computing Environments for Data
      Analysis. Statistical Science, 1, 259-271.

4 An Expert System Approach for Generating and Testing Statistical Hypotheses

K. M. Wittkowski

Department of Medical Biometry, University of Tübingen, Fed. Rep. of Germany

1. INTRODUCTION

The more research aims at combinations of different effects on observational units (people, patients, animals, companies, etc.) and even small differences within subgroups become important, the more results depend on uncontrollable sources of variation within observed data. To cope with this type of uncertainty, a multitude of observational units and complicated statistical models are necessary. Size of the study and complexity of the algorithms require electronic data processing.

The use of common statistical analysis systems (BMDP, P-STAT, SAS, SPSS etc.) has reduced the possibility of numerical errors in statistical analysis. None of these systems, however, can detect semantic errors, because they contain neither (meta-) knowledge on statistical concepts nor knowledge on statistical methods, design of the current experiment or study, primary goal of the analysis, or prior analyses computed on the current set of data. As long as analysis systems were used exclusively by experts in the field of statistics the absence of knowledge to guide the user in selecting methods and interpreting results did not cause serious errors. Because these systems are now relatively easy to be used the need to understand the assumptions implicit to the methods seems less essential so that they are used more frequently also by people with ample expertise in the area of application, but little expertise in the field of statistics (problem-solvers, e.g. physicians, biologists, sociologists). As a consequence, often results based on mis-use of methods and mis-interpretation of "significance" are published e.g. in medical journals (c.f. Altman, 1982). More "intelligent" systems could help the user to avoid at least some of these errors.

Until now, artificial intelligence was mostly used in fields where inexact reasoning is most appropriate. This paper shows how exact reasoning can be used by an expert system to choose an appropriate statistical model given the design of an experiment or study from a more general class of methods and thus avoid certain types of errors in selection of statistical methods and interpretation of statistical results, if structural information corresponding to the experimental design is utilized to simplify both the process of knowledge acquisition and application.

Section 2 illustrates some types of errors that frequently arise from the difficulties with common statistical analysis systems. In section 3 sources for expertise in applied and mathematical statistics are being discussed. Section 4 presents a brief review of expert system approaches in statistical analysis. Section 5 gives a detailed description of the proposed concept of structuring the knowledge needed for statistical analysis. This knowledge is classified in terms of formal and actual relations and problems are classified as conceptual, external, normalized and implicit, respectively. Based on this structured knowledge representation, in section 6, a special pattern-recognition process is introduced and the consequences of this approach on knowledge acquisition and application are discussed. In section 7 statistical knowledge is classified with respect to the stages of the pattern-recognition process described in section 6 and different types of representation are suggested. In section 8, finally, a new type of man computer interfaces is proposed that utilizes and displays structures in the knowledge. Some consequences of these new concepts are outlined in section 9.

2. ERRORS IN STATISTICAL ANALYSIS

Since statistical analysis packages (e.g. BMDP, P-STAT, SAS, SPSS) have become available in almost any research institute, statistical results typically are correct, as far as computational errors are concerned. Statistical results, in spite of their numerical correctness, are often semantically wrong, because the wealth of methods available in modern statistical analysis systems, obviously, often leads to erroneous applications of statistical methods.

Typical examples of errors in using statistical analysis systems are:

- The (implicit) type of question underlying the statistical method chosen differs substantially from the question meaningful in terms of the study. (E.g. a Wilcoxon U-test may be chosen because the data were assumed to be non-normally distributed, but the result is interpreted in terms of "average costs".)

- A paired t-test was chosen although the observational units in both groups are independent, or a high correlation is computed, because the "dependent" variable was originally computed from the "independent" variable.

- A random effect was interpreted as a fixed effect during analysis of variance in a linear model with interaction, or a (first-order) regression was computed to explain a linear relation between interdependent variables.

It is not surprising that the above semantic errors cannot be detected by common statistical analysis systems (Fig. 1): Although the control languages have become simple enough to produce at least some kind of statistical analysis that might look meaningful, the systems still are designed to be used exclusively by experts in applied statistics (Luce, 1980). Consequently, they "cannot store data structures sufficiently" (Haux, 1983b) to provide advice to statistically naive users what type of (implicit) problems the statistical methods have been developed for, i.e. how to select a method or to interpret its "significant" results.

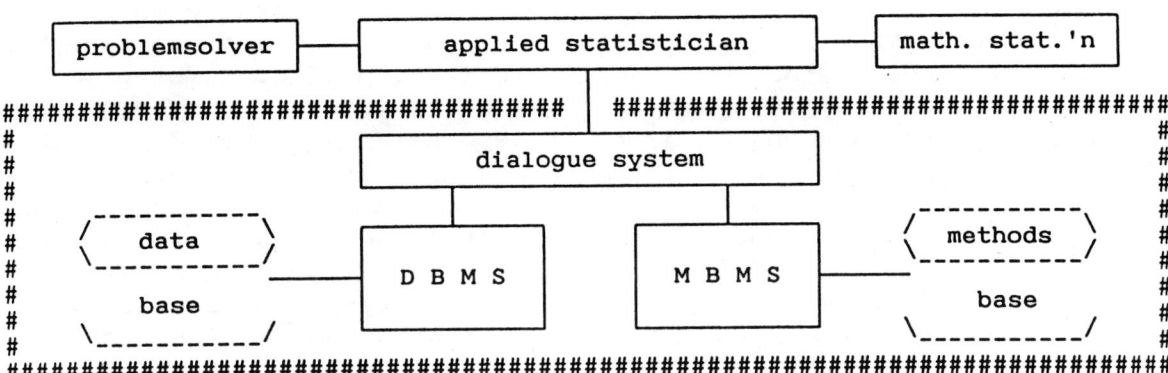

Fig. 1: Components and interfaces of a common statistical analysis system (DBMS/MBMS = data base/method base management system)

3. EXPERTISE IN STATISTICAL ANALYSIS

Recommendations given in documentation or literature, are often not very helpful for scientists with little expertise in the field of statistics. Literature on theoretical aspects of statistical models is typically based on assumptions that are not realistic for many applications: Normal distribution of residuals, for instance, can only be justified if all (unknown) sources of variation are known to be both additive and independent ! As a consequence, the asymptotic relative efficiency (ARE), though important in the field of theoretical statistics, is of little value in the field of applied statistics, especially if sample sizes are relatively small.

Literature on applied statistics and documentation of analysis systems often contain misleading or even wrong recommendations. Conover and Iman (1981), for instance, suggested that analysis of variance methods should be generally applied to "rank transformed" data even in factorial designs. This approach is recommended e.g. in SAS Institute (1984), although already Noether (1981) and Flingner (1981) in their comments on the article by Conover and Iman in the same issue of the American Statistician have proven it to be invalid, because it leads to test statistics where, depending on the ranking procedure, the test on interaction depends on main effects. On the other hand, the assumptions of normality and homoscedasticy of residuals, often taken as reasons for choosing "distribution-free" procedures or transforming the data, are relatively unimportant to analysis of variance procedures (including the well-known t-test).

Transforming the data, however, may lead to a transformation of the problem as well (a t-test on the logarithms of heteroscedastic data, for instance, is sensitive to differences both in location and in scale of the original data). Consider the problem of modelling the expected effect of sales force on sales revenue of Lubinsky and Pregibon (1987) and the following hypothetical data: With a force of 200 or 220 there are observed revenues of 158, 232, and 360 (mean revenue = 250) or 100, 250, and 530 (mean revenue = 293), respectively. In this set of data, an increase in force by 10% increases revenues in the (arithmetic) mean by 12%. Where logarithms

are taken, it is implicitly assumed that an increase in revenue
by 100% in one firm is as relevant as a reduction by 50% in another
firm and the model is based on geometric instead of arithmetic means.
In the current example, geometric means are identical in both cases
so that it might be concluded from the same set of data that
increased force has no effects on revenues at all. Where revenue
is measured in terms of money, the model based on geometric means
is obviously not appropriate and logarithms, consequently, must not
be used, even if the fit is better. As the example demonstrates,
it is not sufficient to pick the pair of transformation functions
$f_i(x)$, $f_j(y)$ that minimizes residuals in the model. Instead, the
choice of transformation should be based on knowledge on the
relevance of differences, i.e. whether an increase by 100% is as
relevant as a reduction by 100% (use the original data) or by 50%
(compute logarithms).

It could be argued that there is no need to find other
representations of statistical expertise, because many erroneous
applications of statistical methods could be avoided if no one but
expert statisticians were allowed access to statistical analysis
systems. Taking into account that many problem-solvers feel the
need to have direct access to their own data and have learned to
handle these systems, at least to some extent, it seems doubtful
that such a regimentation could be enforced. In addition, such
regimentation is not suitable, especially in the field of
exploratory data analysis, where many different sub-problems
(external problems) concerning a given experiment or study
(conceptual problem) are to be analyzed. " Whether we like it
or not, most statistical computations are carried out without
supervision of a professional statistician " (Molenaar, 1984).

4. EXPERT SYSTEMS IN STATISTICAL ANALYSIS

In the Seventies, the first programs for solving problems that
require expert knowledge (expert systems; Buchanan, 1982) were
developed for application in medical diagnosis (i.e. MYCIN, CASNET,
INTERNIST, and PUFF: c.f. Spiegelhalter and Knill-Jones, 1984) and
chemical analysis (DENDRAL: Buchanan and Feigenbaum, 1978). Their
knowledge bases consist mainly of heuristic rules of the form
"IF (condition) THEN (conclusion) WITH (degree of certainty)" and
"probable" conclusions are drawn by means of inexact reasoning.
Only few routine applications of these programs, however, have been
documented, mainly because the lack of meta-knowledge in these
fields results in a difficult and time-consuming process of
knowledge acquisition (Duda and Shortliff, 1983).

Also in the Seventies, attempts were made to define what types of
semantic errors could (and should) be avoided by "more intelligent"
statistical analysis systems. Application of artificial intelligence
to statistics, consequently, was soon introduced e.g. by Hajek and
Ivanek (1982) for hypothesis generation in the field of exploratory
analysis (GUHA 80), by Pregibon and Gale (1984) and Oldford and
Peters (1984) for fitting regression models to observed (inexact)
data, and by Smith, Lee, and Hand (1983) for multivariate analysis.
None of these systems, however, contained enough semantics to
avoid erroneous applications of statistical methods: The purpose

of GUHA 80 was to look for some "significant" correlations in a given set of variables, but not to decide whether or not these correlations were meaningful or to interpret the results to the user; the purpose of REX and DART was to find some transformation that gives the best fit of a regression line to the transformed data, but not to decide, whether or not this transformation was meaningful or to interpret the effects of that transformation on the problem under consideration (see the discussion on transformations above); the area of BUMP was restricted to a relatively small domain of statistical methods.

As in many other areas, where expert systems have been investigated, none of these approaches has been widely accepted, mainly because knowledge acquisition and application proved to be too difficult and time-consuming. REX is being further studied as a component of a new system, called STUDENT (Gale and Pregibon, 1984), currently under development and has influenced the design of the new prototype system TESS (Lubinsky and Pregibon, 1987); Oldford and Peters (1985) dropped their original concept (DART) and started a new approach, called DINDE. More user-friendly procedures for multivariate analysis within common statistical analysis systems have made BUMP superfluous, but it has influenced design of a new series of expert systems (STAT1, STAT2, KENS) that tackle different aspects of statistical analysis using different AI-techniques (Hand, 1987). Most of these systems were never intended to be distributed. Even the knowledge based front-end to GLIM (Nelder and Wolstenholme, 1986), however, is still not ready for distribution. Hand (1985) argues that the reason for the multitude of problems is that both structured and unstructured knowledge have to be taken into consideration, and that these different types of knowledge call for different methods of representation.

5. STRUCTURING KNOWLEDGE FOR STATISTICAL ANALYSIS

Most expert systems have been developed in areas where only few strict rules are known. Heuristic rules and inexact reasoning, consequently, were considered to be most appropriate to handle knowledge in expert systems. Experiences with expert systems show, however, that knowledge acquisition and knowledge application become the less time-consuming the more structure is known in the knowledge, i.e. the more types of objects and relations are known; then it is not necessary to describe each object or relation in detail, but the description of an object or relation can be inherited by the definition of the object- or relation-types it belongs to. Unfortunately, there exists no unique classification of objects and relations. " If ", however, " the statistical community succeeds in producing a workable classification of all or most data sets, a statistical expert system could be very helpful in assessing the adequacy and robustness of some statistical techniques for the particular data set considered. " (Molenaar, 1984)

Looking for such a classification is much more promising in the field of statistical analysis than, for instance, in medical diagnosis: Statistical analysis is predetermined by the way an (prospective) experiment is planned or data are collected in a (retrospective) study. Aside from meta-knowledge of statistical concepts, the set of objects and relations known on an experiment or study and its goals, therefore, might determine a sufficient set of information for an expert system to select an appropriate statistical method, provided the set of methods known to the system is not too extensive. The assumptions implicit to this set of statistical methods determine which objects and relations are necessary.

A concept for structuring knowledge on both statistical problems and methods was introduced by Wittkowski (1984a). It is based on a classification of relations as formal and actual relations. **Formal relations** determine all the knowledge, assumptions, and hypotheses that are independent on the observed data. Additional knowledge based on the observed data is described by **actual relations**. Formal relations can be further classified:

1) A **q u e s t i o n** consists of

   theoretical relations, i.e. non-observable assumptions and knowledge on the reality (SI-units, format of data, strategy of sampling, classification of independent, nuisance, and dependent variables, level and type of scale, theoretical distribution of residuals etc.),

   hypothetical relations, i.e. non-observable, but testable declarations of relevant types of influence on dependent variables (whether there are any differences in distribution, differences in expectation, tendency, dispersion, or whether only linear or monotone differences are of interest), and **output type**, i.e. requirements on the representation of the results (e.g. tabular or graphical representation, computation of a test statistic).

2) A **m o d e l   t y p e** describes the knowledge on **observable relations** within attributes, i.e. all types of observational units with respect to the hierarchy of the independent variables (all types of relation) and the number of tuples for each type of relation.

The knowledge represented within the **d a t a** on **observed relations** selects an actual relation for each type of relation. A combination of question and model type will be referred to as **p r o b l e m   t y p e** , model type and data as **m o d e l** , and all three, question, model type, and data as **p r o b l e m** .

The distinction between actual and formal relations is essential to the concept of knowledge engineering: Actual semantic integrity constraints can be checked either by data base management systems (this feature is to some extent already implemented within some systems) or inside the method (cf. Nelder, 1977). Little attention, however, has been given to formal semantic integrity constraints (Haux, 1983a).

As proven in Wittkowski (1986), formal relations are sufficient for choosing appropriate statistical methods and interpreting their results by means of artificial intelligence techniques, as far as consistency is concerned, provided that a suitable class of statistical methods is selected: i.e. linear models (analysis of variance and covariance), tendency models (several non-parametric models based on ranks), "semi"-parametric models (ranking after alignment), log-linear models (analysis of contingency tables),

or several graphical and tabular techniques. It can easily be
generalized e.g. to principal components analysis, analysis of
dispersion etc. .

A set of objects and relations, describing all the knowledge and
assumptions known prior to the collection of the data, and also the
goal of the experiment or study will be called the **conceptual
problem type**. (The terms **conceptual question, conceptual problem,
conceptual model type,** **conceptual model**, and **conceptual data** will
be used accordingly.) The conceptual problem type defines all
meaningful subproblems that can be defined for a given study, and
the conceptual model type defines the finest possible lattice
structure for the conceptual data (see the first column in Fig. 2).

A sub-problem is typically given by selecting a subset (restriction)
of observational units, selecting a subset (projection) of
interesting variables, and (re)defining the output type. The terms
'restriction' and 'projection' are chosen in accordance with their
use in relational data models. In the field of hypothesis generation
(exploratory data analysis) it might even be necessary to modify
theoretical, hypothetical, or observable relations. In any case,
the expert system can tell the user about possible modifications,
consider the effects on global error rates, and explain the
consequences on interpretation. The sub-problem is sufficiently
defined by the **external problem type** (see the second column in
Fig. 2).

Since formal relations are sufficient for the selection of a
statistical method, a statistical method can be sufficiently
described by the set of all problem types for which this method
is appropriate. These problem types associated with a method are
called **implicit problem types** (see the fourth column in Fig. 2).

Several experimental designs may be logically equivalent: If the
independent attributes are permuted in a hierarchical design, for
instance, the external designs are formally different, although
appropriate statistical methods are the same for all permutations.
To reduce the number of implicit problem types that have to be
acquired for a statistical method, **normalized problem types** are
introduced as representatives of classes of logically equivalent
external problem types (see the third column of Fig. 2).

6. ACQUIRING AND APPLYING KNOWLEDGE IN STATISTICAL ANALYSIS

As a consequence, choosing a statistical method reduces to a
special pattern-recognition process (c.f. Wittkowski, 1985 for
details) which consists of

1) **representing problems** by their conceptual problem type and the
   (conceptual) data (top and bottom of Fig. 2, respectively),

2) **representing methods** by their implicit problem types (top of
   Fig. 2),

3) choosing an (external) subdesign by (**modifying,**) **projecting**, and
   restricting the conceptual problem type (1st/2nd column),

4) **normalizing** the resulting (external) problem (2nd/3rd column),

5) **selecting** a method with a corresponding implicit problem type (top of 3rd/4th column), and

6) **verifying** assumptions of the method on the data (bottom of 3rd/4th column).

Representation of program code (including implicit assumptions), implicit problem type, conceptual problem type, and data are typically defined by a programmer, theoretical statistician, applied statistician and data logger, respectively, before the experiment is started. The problem-solver, typically, enters the set of variables and the type of output he is interested in. All additional theoretical, hypothetical, observable, and observed relations can be derived (by the expert system) from the conceptual question and a corresponding statistical method can be automatically called. The complete pattern-recognition process is given in Fig. 2.

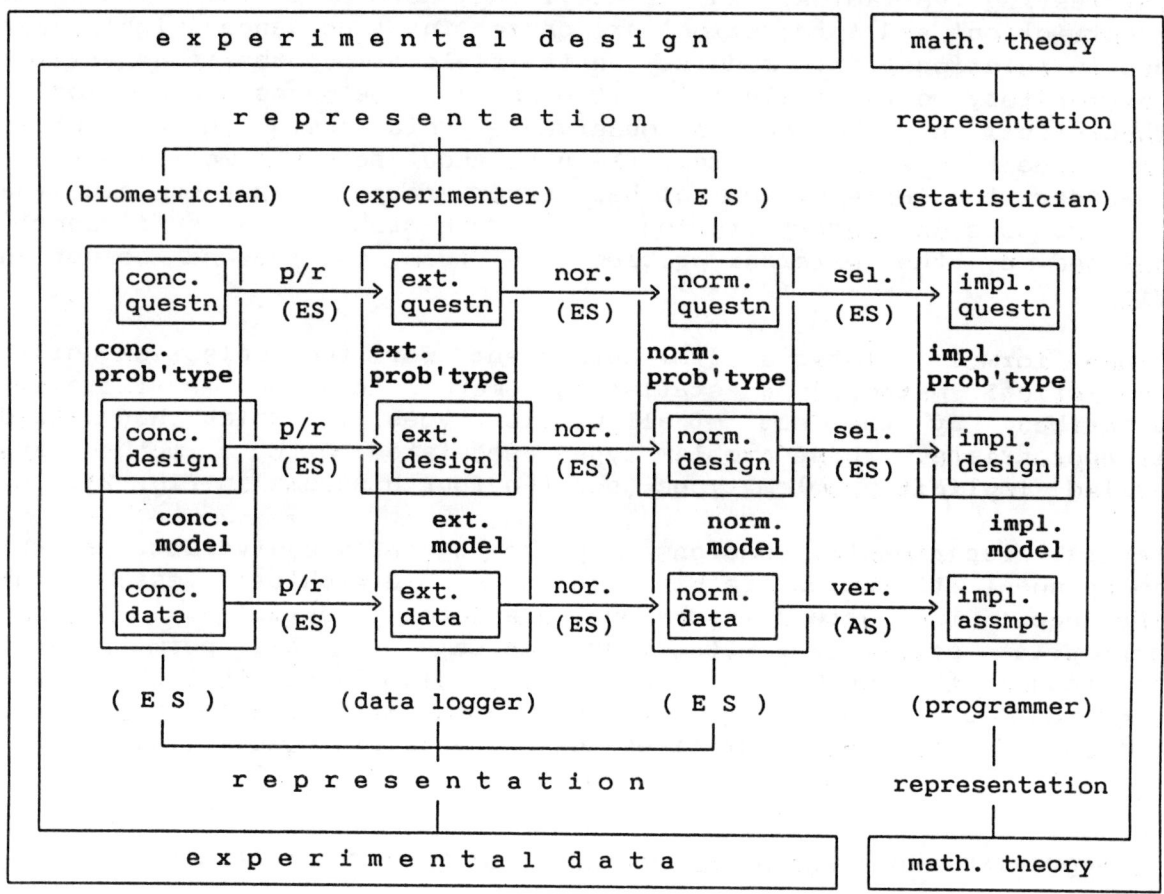

Fig. 2: Choosing statistical methods as a pattern-recognition process (conc. = conceptual, ext. = external, norm. = normalized, impl. = implicit; questn = question, prob'type = problem type, assmpt = assumptions; p/r = projection and restriction, nor. = normalization, sel. = selection, ver. = verification, ES = expert system, AS = analysis system)

Representation of experimental design or mathematical theory cannot be validated by an expert system, unless it contains knowledge on the field where the statistical methods are applied to or can compare a method with its implicit problem types, respectively.

Implementing knowledge on the domain of application was proposed e.g. by Victor (1984), whether or not a description of a method is correct could be tested by an additional component of an expert system by means of simulation techniques, and correctness of a statistical method could possibly be enhanced by the concept of "stepwise reduction of data structures" presented in Haux (1983a). Because the present paper concentrates on testing and generating statistical hypotheses, these aspects are not considered here.

In the pattern-recognition process described above (Fig. 2) knowledge has to be entered and can be used at different stages:

1) Acquisition of knowledge on conceptual problem types can be verified by testing its consistency. Since this type of knowledge is typically entered by an expert in statistics or by a relatively experienced problem-solver, fast dialogue procedures are most appropriate, and, consequently, little redundancy is available to test consistency during definition of conceptual problem types.

2) Acquisition of knowledge on implicit problem types will be handled in an even more formalized manner. This kind of knowledge acquisition is done relatively seldom and always by experts in mathematical statistics.

3) Acquisition of knowledge on explicit problem types can also be tested on completeness. The amount of necessary input from the problem-solver, however, can be reduced by deducing non-modifiable objects and relations from knowledge of the conceptual problem type, and the reduced amount of necessary input can be verified by testing its consistency with the conceptual problem type. The expert system explains all inconsistencies or, alternatively, presents small pop-up menus that contain only those parameters that have already been verified to be consistent alternatives. Because this part of the system will also be used by the problem-solver, it is possible that the user does not even have sufficient knowledge on statistical concepts to choose from the set of parameters presented as meaningful alternatives. In that case, the system teaches the necessary concepts in terms of the conceptual and external problem type by means of intelligent tutoring.

4) Data entered by a data-logger or the problem-solver can be compared with integrity constraints in the current external model type and all relevant integrity constraints of the conceptual model type.

5) In order to choose an appropriate statistical method for the given external problem, the expert system normalizes the external problem type and selects a method with a corresponding implicit problem type.

6) After the method has been selected, the expert system arranges the data and generates the necessary command sequence, so that the appropriate method can be automatically called without the need of further input from the problem-solver. The method itself verifies its assumptions on the (normalized) data.

7) In a similar manner, the analysis system's output can be edited by the expert system in order to eliminate answers to irrelevant questions (e.g. means and standard deviations of nominally scaled data), to re-express its numerical results in the terminology used in the external problem type, and to adapt the results (e.g. of multiple analyses) to the conceptual problem type.

7. REPRESENTING KNOWLEDGE FOR STATISTICAL ANALYSIS

The information necessary at different stages of the above described pattern-recognition process can be used to differentiate several types of knowledge and to suggest appropriate types of representation for each type of knowledge:

1) Knowledge on conceptual problem types (lattice structure of observational units, theoretical and hypothetical relations, output types) should be handled separately (in the **'problem knowledge base'** of Fig. 3), because it depends on the current experiment or study. A declarative representation (in the form of frames) seems to be most appropriate.

2) Knowledge on completeness and consistency of problem types and knowledge on deduction and verification of external problem types (the **meta-knowledge base** of Fig. 3), is independent of both the knowledge on the current experiment and of the methods available. A procedural representation (in the form of rules) seems to be most appropriate for this knowledge base.

3) For the purpose of "intelligent" tutoring, additional knowledge is necessary, describing the strategies of teaching statistical concepts (the **tutor knowledge base** of Fig. 3).

4) The necessity of normalization has already been discussed in section 5. Because representation of knowledge on statistical methods depends on the type of normalization considered, the rules of normalization have to be defined together with the statistical methods. Even if the number of implicit problem types is reduced by normalization, a statistical method, typically, is valid for a multitude of (normalized) external problem types. On the other hand, less object-types might be relevant for implicit problem types than for explicit problem types (SI-units and ranges, for instance, can be neglected) or can be deduced from the rules of consistency. (If there is a rule in (2) that a test on expectation requires a non-nominal scale level, this restriction need not be explicitly given in the implicit problem type of the t-test.) In Wittkowski (1985) objects within several object types are (partially) ordered and a declarative representation of implicit problem types as rectangular sub-spaces of a high-dimensional parameter space is given. For each method, secondly, knowledge about calling that method (arranging the data and choosing parameters in the analysis system's control language) has to be defined. If statistical analysis systems like BMDP, P-STAT, SAS, SPSS are to be called, simple production rules (or even conventional techniques as used in BUMP) seem to be appropriate to describe how to call a statistical method. Both the implicit problem type and these production rules are part of the **'methods knowledge base'** of Fig. 3. For the purpose of interpretation this knowledge base should also include a description, where relevant results can be found in the analysis system's output.

Because these different types of knowledge representation can be associated with different steps in the problem solving process, this approach suggests that not a single system but rather a collection of cooperating systems is appropriate for statistical expert systems.

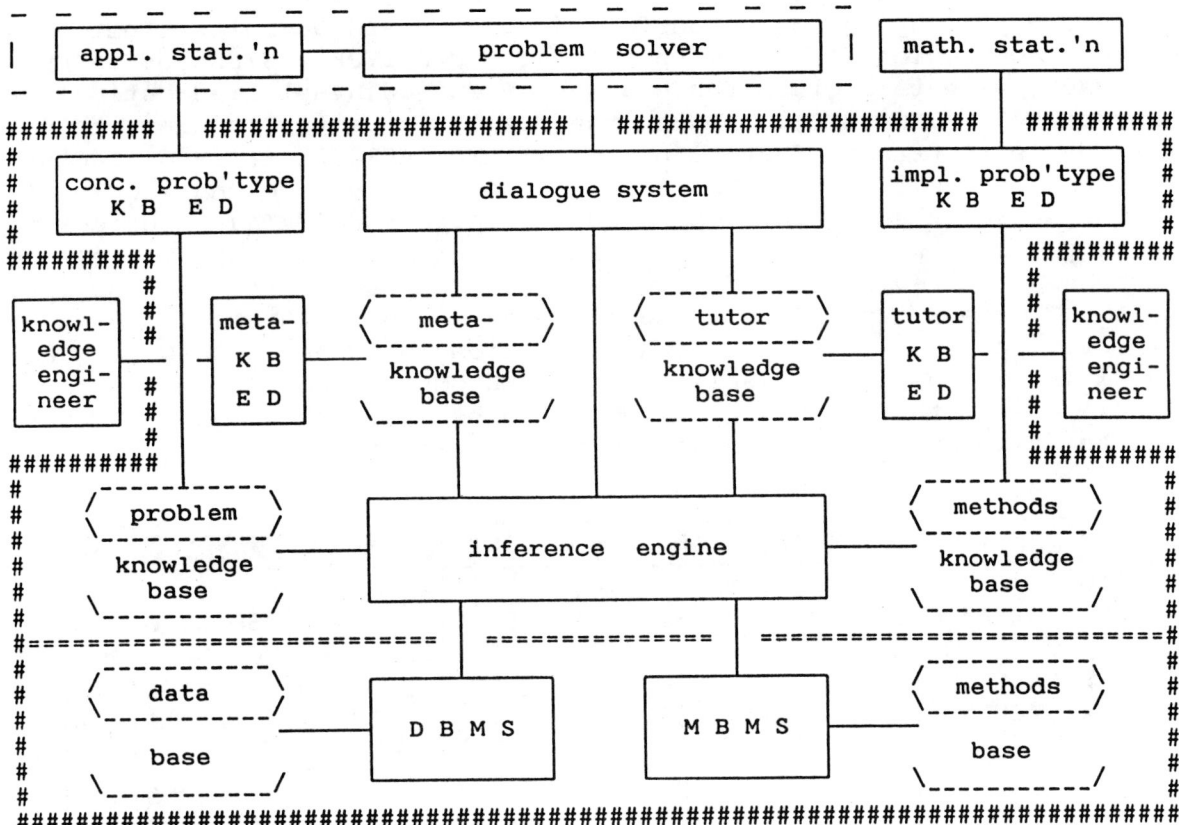

Fig. 3: Components and interfaces of an expert system for statistical applications (KBED = Knowledge base editor). The dashed box indicates that — with help from the expert system based on the tutor knowledge base — part of the activity of a biometrician (appl.statistician) during acquisition of knowledge on conceptual problem types might be taken over by an experienced problem solver.

8. THE USER INTERFACE

Most early expert systems used some kind of restricted "natural" language interface to communicate with the user. These dialogue procedures seemed to be most appropriate for programs that need to handle heuristics by means of inexact reasoning in a domain, where little is known on the structure of knowledge. In most cases, however, this type of dialogue has not been accepted by the user, mainly because natural language tends to be relatively time-consuming. As a consequence, more structured dialogue forms have been considered even for expert systems in medical diagnosis (e.g. ONCOCIN). The concept of structuring knowledge presented in the previous sections are displayed in form of a structured spread-sheet. Most interaction is necessary during definition of external problems and input of (external) data (see 2nd column of Fig. 2.). The following example demonstrates, how the amount of input necessary to define a sub-problem is reduced.

Consider a medical experiment (see Wittkowski 1985 for details), where a bypass operation is performed on 10 cardiac infarct patients either conventionally or with a laser (OPERTYPE = STEEL or LASER) and each patient is given three doses of an adjuvant medication of propranolol (BETADOSE). Thus the design is partially hierarchical with patients (random factor) nested within OPERTYPE (fixed factor) and crossed with BETADOSE (fixed factor). Bodyweight (BODYWGHT) was measured at entry in the study, vigor (ERGOMETR), subjective ranking of well-being (SUBJRANK), and type of heart failure (RHTHMERR) for each dose of propranolol. Suppose that the primary goal of the study was to measure the effect on ERGOMETR in terms of expectation and that a linear relation between BODYWGHT and ERGOMETR is assumed. Selection of "ERGOMETR" and "SUBJRANK" to be displayed and the effects on ERGOMETR to be "TEST"-ed would result in the following (partially reproduced) screen form:

NAM	OPERTYPE	*******	BETADOSE	BODYWGHT	**ERGOMETR**	**SUBJRANK**	NAM
SIU	–	– (1)	mg	kg	kgm2/s2	–	SIU
N/F	002 A5.0	010 N1.0	003 N4.2	010 N5.2	030 N5.0	030 N1.0	N/F
SCL	002 T ND	000 T ND	003 T AC	000 S AC	000 O AC	003 O OD	SCL
MIN	LASER (2)	1	(2) 10.0	(3) 30.0	(3) 0.0	(3) 1	MIN
MAX	STEEL	5	50.0	250.0	500.0	3	MAX
DAT	LASER	1	10.0	74.0	172	1	DAT
DAT	:		20.0	:	190	3	DAT
EDT	:	1	50.0	**74.0**	195	2	EDT
DAT	:	2	10.0	76.0	156	1	DAT
DAT	:		20.0	:	145	2	DAT
I/O	EXPECT	–	EXPECT	LINEAR	**TEST**	–	I/O

Fig. 4: Partial display of PANOS-ES screen form during definition of external problems. (NAM: name of variable, SIU: SI-unit, N/F: no. of observational units/format of data, SCL: size of population/treatment, strata or observation/level and type of scale, MIN/MAX: range, DAT/EDT: data/edit line, I/O: input or output type; **selected** or **changeable** field; (n): notes)

Selecting the items ERGOMETR and TEST is sufficient to enable the expert system to call the appropriate analysis of covariance. Implicit factors (1) are generated according to the hierarchy of the selected variables. If OPERTYPE or BETADOSE are restricted to one or two categories (2), the necessary modifications of the analysis system's control language are automatically generated and the user is given the information, how to compute a global p-value from the local p-value given by the analysis system. Future back-end modules will be able to take over these activities while automatically editing the output. If BODYWGHT, ERGOMETR, or SUBJRANK are restricted (3), the influence types are modified (e.g. a LINEAR relationship is assumed between BETADOSE and ERGOMETR), or a variable PATINGRP with categories 1-5 is introduced (1), the appropriate methods are called as well, but it is pointed out that the result has to be interpreted as being exploratory.

8. CONCLUSIONS

The strategies for knowledge representation proposed in the present paper provide a concept for building expert systems both in generating and testing statistical hypotheses (exploratory and confirmatory data analysis). They allow for designing differently structured knowledge bases and specialized man-computer interfaces.

The advantages of the proposed expert system approach to statistical analysis can be summarized as follows: Utilizing knowledge of the conceptual problem type, defined typically by a biometrician, the amount of information to be entered by an experimenter during analysis of subproblems is significantly reduced. This reduction of input allows for short formalized dialogue procedures and results in fewer erroneous applications of statistical methods. Instead of choosing certain test procedures and defining a complete subset of relevant information for each sub-problem, the experimenter can concentrate on the formulation of questions (defining a projection on the data set and hypothetical relations). In exploratory data analysis, the set of applicable methods is reduced to methods that reflect the theoretical and observable relations of the study and the experimenter is given results in terms of hypothetical relations. Thus, statistical results will become easier to interpret and to survey, so that even seemingly contradictory results (significant vs. non-significant) might complement one another.

As far as implementation is concerned, the separation of data and methods vs. knowledge on data, methods, and concepts allows for a separation of the expert system from the data base management and statistical analysis systems. As a consequence, the expert system may be implemented on a PC even if the data set is large. The same analysis systems may be utilized for routine applications (with the expert system as interface) and special applications (directly by the biometrician). Especially if data and knowledge are stored on distant computer systems, separation of data and knowledge provides for more protection against unauthorized access to the data (Wittkowski, 1984b), because data (typically stored and evaluated on a host) is meaningless without the semantics stored within the knowledge bases (on a remote PC).

Thus, finding a "workable classification" (c.f. section 5) not only simplifies implementation, knowledge acquisition, and knowledge application by allowing for a simpler control strategy in the process of deduction and faster dialogue procedures, but also gives more insight to common concepts underlying statistical methods. This will lead to two more consequences: The concepts will become explicitly defined, so that they can be discussed among experts, misleading or wrong heuristic rules can be corrected, and the statistical results will become less dependent on the subjective opinion of a special expert. The second consequence is related to teaching statistical concepts: The experimenter will be taught concepts instead of special methods so that teaching can concentrate on the essential process of formulating problems and results in terms of hypothetical relations, while unnecessary technical details can be omitted.

ACKNOWLEDGEMENTS

Part of this paper is based on the chapter "Generating and testing statistical hypotheses: Strategies for knowledge engineering" in Haux (ed.) "Expert systems in Statistics" Gustav Fischer Verlag Stuttgart, New York 1986.

REFERENCES

[1]   Altman, D.G. (1982),   Statistics in medical journals. Statistics in Medicine 1, 59-71.

[2]   Buchanan, B.G. (1982),   New research on expert systems. Machine Intelligence 10, 269-300.

[3]   Buchanan, B.G. and Feigenbaum, E.A. (1978), DENRAL and META-DENDRAL, their application dimensions. Artificial Intelligence 11, 5-24.

[4]   Conover, W.J. and Iman, R.L. (1981), Rank transformations as a bridge between parametric and nonparametric statistics. The American Statistician 35, 124-133.

[5]   Duda, R.O. and Shortliff, E.H. (1983), Expert systems research. Science 220, 261-268.

[6]   Flingner, M.A. (1981), Comment on a paper by Conover and Iman. The American Statistician 35, 131-132.

[7]   Gale, W.A. and Pregibon, D. (1984), Constructing an expert system for data analysis by working examples. In: Havranek, T., Sid k, Z., and Novak, M. (ed.), COMPSTAT 84. Wien: Physica, 227-236.

[8]   Hajek, P. and Ivanek, J. (1982), Artificial intelligence and data analysis. In: Caussinus, H., Ettinger, P., and Tomassone, R. (ed.), COMPSTAT 1982 - Part I, proceedings in computational statistics. Wien: Physica, 54-60.

[9]   Hand, D.J. (1985), More intelligent statistical software and statistical expert systems. The American Statistician 29, 1-16.

[10] Hand, D.J. (1987), The application of expert systems in statistics. (present issue)

[11] Haux, R. (1983a), Die Verwendung komplexer Datenstrukturen in statistischen Auswertungssystemen. Ulm, F.R.G.: Dissertation.

[12] Haux, R. (1983b), How to detect and prevent errors in computer-supported statistical analysis, an example. Methods of Information in Medicine 22, 87-92.

[13] Luce, S.R. (1980), A conceptual analysis of SPSS and BMDP. In: Barritt, M.M. and Wishart, D. (ed.), COMPSTAT 80. Wien: Physica, 509-514.

[14] Lubinsky, D. and Pregibon, D. (1987), Data analysis as search. (present issue)

[15] Molenaar, I.W. (1984), Discussion on the paper by professor Victor. Statistical Software Newsletter 10, 121-122.

[16] Nelder, J.A. (1977), Intelligent programs, the next stage in statistical computing. In: Barra, J.R., Brodeau, F., Romier, G., and Van Cutsem, B. (ed.), Recent developments in statistics. Amsterdam: North-Holland, 79-86.

[17] Nelder, J.A. and Wolstenholme, D.E. (1986) A front end for GLIM. In: Haux, R. (ed.) Expert systems in Statistics. Stuttgart, New York: Fischer. 155-177

[18] Noether, G.E. (1981), Comment on a paper by Conover and Iman. The American Statistician 35, 131-132.

[19] Oldford, R.W. and Peters, S.C. (1984), Building a statistical knowledge based system with Mini-MYCIN. Proc ASA Statistical Computing Section, 85-90

[20] Oldford, R.W. and Peters, S.C. (1986), Statistically sophisticated software and DINDE. Workshop on Artificial Intelligence and Statistics, 11-13 April 1985, Princeton, NJ.

[21] Pregibon, D. and Gale, W.A. (1984), REX: an expert system for regression analysis. In: Havranek, T., Sidak, Z., and Novak, M. (ed.), COMPSTAT 84. Wien: Physica, 242-248.

[22] SAS Institute (1984), SAS user's guide: statistics. Cary, NC: SAS Institute Inc.

[23] Smith, A.M.R., Lee, L.,S., and Hand, D.J. (1983) Interactive user-friendly interfaces to statistical packages. The Computer Journal 25, 199-204.

[24] Spiegelhalter, D.J. and Knill-Jones, R.P. (1984), Statistical and knowledge-based approaches to clinical decision-support systems, with an application in gastroenterology. Journal of the Royal Statistical Society, Series A 147, 35-77.

[25] Victor, N. (1984), Computational statistics: tool or science ? - Werkzeug oder Wissenschaft ?. Statistical Software Newsletter 10, 105-116.

[26] Wittkowski, K.M. (1984a), On the use of structural information for a statistical expert system in medical research. In: Van Eimeren, W., Engelbrecht, R. and Flagle, C.D. (ed.), Third international conference on system science in health care. Berlin: Springer, 1140-1143.

[27] Wittkowski, K.M. (1984b), Künstliche Intelligenz und verteilte Intelligenz: Implikationen für Sicherheit, Interpretation und Schutz von Krankendaten im Rahmen klinischer Studien. In: Abt, K., Giere, W., and Leiber, B. (ed.), Krankendaten, Krankheitsregister, Datenschutz. Berlin: Springer, 145-149.

[28] Wittkowski, K.M. (1985), Ein Expertsystem zur Datenhaltung und Methodenauswahl für statistische Anwendungen. Stuttgart, F.R.G.: Ph.D. Dissertation.

[29] Wittkowski, K.M. (1986), An expert system for testing statistical hypotheses. In: Boardman, T.J. (ed.) Computer Science and Statistics: Proceedings of the 18th Symposium on the Interface. Washington, D.C.: American Statistical Association, 438-443.

5 Dialogue Management with Computer-Based Statistical Analysis

S. M. Furner
British Telecom, Human Factors Division

INTRODUCTION

The technology of statistical analysis is becoming more accessible within a commercial environment due to its increasing availability in the form of computer software. The computational power of the machine allows the user to carry out statistical analyses which would have been prohibitively costly in effort and time if conventional hand calculation were employed. While these tools have become more common, fears have been expressed that adequate training in the skills required to reliably use them has not. As a consequence it has been suggested that the software to conduct statistical analyses should contain an expert system knowledgable about the statistical procedures the software provides. [1]

The integration of AI techniques within statistical software has tended to concentrate upon the problems of adequately capturing statistical skills within the knowledge base of an expert system. Typically only one expert system is envisaged as being associated with a statistical software package. [2] [3] [4] However, this may only be one aspect of the role AI could be expected to play in commercial packages intended for industrial use.

I would like to argue that there is a role for the serious consideration of the needs of the consumer of statistical technology, and the practical constraints he or she may be facing in making use of statistical software. Pragmatic solutions adopted by the statistical consumer should not result in a degradation of the quality of the analysis being undertaken. The use of knowledge-based programming techniques may provide a means of aiding the statistical consumer in exploiting the tools available to produce the optimum solution for his or her situation. To achieve this there may be more than one area within a statistical software package which makes use of knowledge-based engineering techniques.

ANALYSIS IN PRACTICE

A large multi-purpose statistical tool such as SPSS or SAS, both of which may be used commercially in a range of industrial settings, currently require more than a knowledge of statistics to be effectively used. In carrying out even the simplest analysis the data must first be entered into the package. The package's user interface has to be mastered in order to build a data-set and specify the analysis to be undertaken. There must also be sufficient computing resources available, within the machine on which the package is mounted, to allow the software to successfully carry out the calculation. It is not unknown for researchers to work with SPSS only at night, because it is the only time sufficient machine resources become available for a specific statistical procedure to be carried out. [5]

Possibly, one of the key skills required by the practical user of statistical software is the ability to accurately interpret an error message. Within an interactive session with a mainframe computer these can come from more than one source. They may be produced by the package being used or be presented by the operating system within which the package runs. If a micro-computer is being used as a terminal then they may originate from the terminal emulation software, or a file transfer program. If the connection between the terminal and the mainframe is via a local area network, then this too may produce its own crop of messages.

An error message, which was experienced by an Ergonomist I was assisting to analyse experimental data recently, was "ABEND ERROR" followed by a numeric code. This message provided no meaningful information to myself or the individual working directly with the data. It was necessary for us to contact the computer centre where the machine we were using was located. Here we found a skilled programmer sufficiently familiar with the operating system to be able to interpret the message for us. According to the programmer the statistical package had called a program from a general purpose library maintained on the computer, this program had failed to run properly. As a result of experience gained in using the system she suggested that this probably resulted from insufficient working memory being available for use by the library program. In plain English, the analysis failed because there was insufficient machine resources available to the statistical package at the time the analysis was run.

Two points need to be made here. Firstly, and most obviously, the error message provided insufficient information for the computer user who received it. This resulted in delays to the analysis while the user located a specialist to interpret the message and take action to deal with the problem. From the programmer's response to the query this was not an isolated occurrence.

The second point that the error message raises is that practical problems of implementing an analysis are not independent of the choice of statistical procedures which are employed. Once the problem had been identified as one of machine resource limitation, two routes were open to deal with it. The first was to obtain an increase in machine resources. The second was to carry out the analysis using statistical procedures which required less machine resources.

In connection with this problem it should be noted that the selection of the appropriate solution can be confused by apparently helpful error messages. The content of an error message may have consequences for the way in which a machine problem is dealt with. For example, a statistical package may fail to run a calculation and return an error message indicating that there was insufficient working memory available. The message could contain a suggestion of the working memory

size which should be available for the analysis. After contacting the
computer centre and arranging an increase in working memory allocation,
the analysis could again fail to run due to insufficient memory.

On further investigation of the software package it may be found that
the working memory size suggested by the error message was the memory
size at which the program had failed to run plus an arbitrary increase.
The value of the increase in this situation would not be dependent upon
the requirements of the program to perform the calculation. It is very
easy for the unwary to be fooled into false conclusions of accuracy by
the apparent precision of a clearly specified numeric value. Error
information needs to provide a clear unambiguous indication of the
nature of the problem being indicated.

What can be concluded from this problem with error messages? Most
obviously, that the practical implementation of a computer based
analysis can incur problems of software usability, even for those
packages which it could be considered easy to employ. General purpose
statistical tools such as SPSS, SAS, or other packages currently
available, are easy to use in comparison to the tools which were on
offer before them, however, this does not mean that they are without
difficulty for the user.

The less obvious, and most important conclusion, is that practical
problems of running statistical software interacts with a user's
statistical knowledge to influence the strategy employed in the
analysis. For the purpose of a practical statistical analysis a source
of expertise can be identified as being required. This is a knowledge
of the characteristics of the computing system being used, and the
limitations of its operation. The next section discusses how this
knowledge source could be integrated into the structure of a user
interface.

DIALOGUE MANAGEMENT

Conventional software architecture results in an interaction between the
user and the program being generated from different sections of the
program code. Calls are made to the user's terminal from within the
program code when information is required, or is to be displayed. This
approach can result in a dialogue which lacks flexibility and may be
inconsistent in its presentation. Rather than allow the user to
interact directly with the different sections of the program code, part
of the program can be specifically written to support the dialogue.
This would be an independent unit placed between the user and the
program application code. The technique of producing an independent
section of the code, specifically dedicated to organising the flow of
information between the user and the application code, is known as
dialogue management.

With the flow of control of the user dialogue no longer embedded within
the application code its design and implementation becomes more easily
organised. Since the user interface is now independent of the
application code, it can be constructed and tested separately. Thus,
the production of alternative prototype dialogue systems becomes
possible for a specific application. These alternative prototypes can be
constructed and evaluated before, during, or after the application code.

A dialogue manager will organise the style and content of the interaction between the application code and its user. The aim of the manager is to minimise the difficulty experienced by an individual in carrying out his or her goals in using the software. The presentation of the software at the terminal may be designed to produce a consistent metaphor, or image, to suit the domain knowledge of the user population. A specific style of interaction may be implemented by the manager and consistently used for all interactions with the user.

An example of a consistent style would be the use of a formatted mode of entry for the presentation of a system to its user population. This style is usually applied where a system is to be used on a casual basis, or by staff not skilled in computing. The user is only allowed limited communication with the software and each transaction is constrained by the program to a predetermined range of noncatastrophic outcomes. Essentially, the substance of the interaction between the machine and the user population consists of the presentation of screens of prepared information to which the user can respond by filling in blank slots. It is very similar to the presentation of a series of paper forms, when one is filled in the user goes on to the next. In this way information can be quickly obtained or provided.

There are limitations upon the types of problems a user may experience, that may be controlled by a dialogue manager. For example, system response time can significantly influence user attitude and performance. [6] This type of dialogue problem cannot be eliminated by dialogue management. However, the manager is in a position to reduce the impact of the length or variation of the response time upon the user. Warnings can be given if an activity is going to result in a prolonged delay. An accurate assessment of the delay length, with an indication of the time until completion of the task, may help to reduce the effect upon the user by freeing him or her from the terminal to carry out other tasks. [7]

INTELLIGENT DIALOGUE MANAGEMENT

The separation between the dialogue control and the application code allows specialist techniques and software tools to be introduced into the design of the interface. IBM currently provides a proprietary dialogue management tool called the Interactive Systems Productivity Facility (ISPF). [8] This is based around the provision of screens of information and contains facilities for their specification and interaction with the application code. ISPF dialogues tend to make use of a menu based interaction with the user. The ISPF dialogue is essentially an add on to the application which makes use of conventional programming techniques to organise the dialogue with the user.

The ISPF can be contrasted with a more sophisticated management tool such as SYNICS produced by Leicester Polytechnic. [9] This is an intelligent system in that it contains provision for rule based specifications to be included within the interface. A SYNICS dialogue is intended not only to organise the interaction between the user and the application code, but also to use the rule base to translate inputs from the user or the application. A SYNICS interface can use its rule system to parse an input and make a decision about what action it should take to deal with it.

A SYNICS dialogue manager makes use of the knowledge within its rule set to transform incomprehensible output from the application into a form meaningful to the user. Conversely, it takes users input and employs the knowledge in its rule set to change it into a form which is comprehensible to the host application. By this method conventional restrictions upon the form of user input to an interface are made much more flexible. The use of knowledge-based engineering techniques makes it possible to consider very loosely specified user input to a program, which could take the form of near natural language requests.

Prototype interfaces have been produced in SYNICS to investigate its ability to deal with natural language specifications. [10] The domain in which the prototypes operated was requests for train time-table information. The prototype was constructed to function as a buffer between a speech recognition system and a data-base of train time-table information. (In fact, neither the speech recognition system nor the data-base existed.) The prototypes ran under the control of a test-bed program, which enabled it to run without a host application. Test sentences could then be presented to the prototype and its behavior observed.

The prototype was able to parse a sentence to discover if there was sufficient information for a time-table specification. If there was not a full specification it would look for a partial specification and then enter a limited dialogue to recover the missing information from the user. If it could find neither a full or a partial specification it would look for clues within the sentence that might indicate elements of a time-table specification. Although not implemented within the prototypes, the intention was that information gained by this last strategy could be used as the basis for a dialogue with the user to obtain a full specification.

The rules which were used within the prototype were obtained from experimental data of live speakers. Previously, an experiment had been carried out to investigate users' reactions to different voice types used to provide spoken prompts in a simulated voice recognition system. The requests for information were freely produced, unconstrained responses by the subjects. The text of the requests produced by the subjects for the natural voice condition of the experiment, were analysed to establish semantic definitions of the content of a request for the rule set in the prototype interfaces.

While interface systems such as the prototype train time-table system can appear to provide unlimited natural language capacity in their presentation to the user, they operate within a limited natural language domain, and are restricted in the types of natural language constructions which can be dealt with. The train time-table program made no use of syntactic knowledge, it relied solely on a series of semantic definitions of the content of a legitimate natural language time-table request. Constraints upon vocabulary, grammar and semantics act to limit the the range of specifications a user can employ when interacting with a near-natural language system. The limited near-natural language available in this situation is sometimes known as a "fortual" language. [11] This is a derivative of the description formal-natural language, a language which should possess the following properties:-

a) It is a subset of a natural language.

b) It is designed to be easily learned by the natural language speakers.

c) It is embedded in a system organised to defend against the hazards of vagueness and ambiguity usually associated with natural language.

For applications in a statistical interface, dialogue management of the variety offered by tools such as SYNICS, could provide very flexible methods of specification of the statistical procedures to be carried out. By virtue of the knowledge encoded in the rule set, domain dependent specifications could be interpreted into accurate statistical specifications.

The addition of knowledge-based engineering techniques to provide flexibility for the user input, may reduce difficulties of learning the command structures employed by a statistical software system. The method of specifying the analysis would be the commonly used terms and methods of explanation from the domain in which the user operated. With the addition of another source of knowledge to deal with user communication, the structure of an intelligent interface begins to take on the form of a collection of knowledge-based systems, rather than the addition of a single statistically knowledgable expert system.

THE CONTENT OF AN INTELLIGENT INTERFACE

To effectively support and manage a user's interactions with an application, Rissland [12] identifies seven sources of knowledge which would be included in an intelligent interface. These are:-

1) Knowledge of the user
2) Knowledge of user tasks
3) Knowledge of tools
4) Domain specific knowledge
5) Knowledge of interaction modalities
6) Knowledge of how to interact
7) Evaluation knowledge

For a statistical system the statistical knowledge would be primarily maintained within (4) as the "Domain specific knowledge" of the interface, although it may also appear in (2) the "Knowledge of user tasks". (3) the "Knowledge of tools" refers to the capabilities of the machine being used. It would deal with machine resources, protocols etc.; it is here that the ability to accurately define the cause and implications of a system error would be located. The presentation of the system to the user would result from (5) the "Knowledge of interaction modalities" and (6) the "Knowledge of how to interact". The ability of the interface to cope with differing styles of user interaction would be determined by the scope of the knowledge contained within sources (5) and (6).

The integration of the knowledge sources into the interface can result in their maintaining an individual identity, or they could be subsumed within the sub-systems of the interface structure. An object-oriented programming environment supports agents which are highly modular and independent, which communicate both among themselves and with any human participants. The structure of the interface in this situation would be designed to organise the agents' capabilities and communication efficiently. A basic outline of the advantages of the object-oriented approach for statistical data analysis, and a description of a basic system are discussed by Deken [13]. Deken's system employs three simple agents. One to capture the data, another to organise the interaction between the agents, and a third to report on the state of the analysis and the interaction between the other agents.

A more conventional approach to the problem of intelligent interface architecture is taken by Kuo and Konsynski [14]. They suggest the use of 4 functional layers between the application and the user. The functions in each layer are free to perform without concern for the internal procedures of the other layers. Clearly specified protocols are used to communicate between layers of the interface.

The first layer is the "environmental interface", this deals with the terminal hardware. It collects the user input from the different types of input device that are used, and presents the interface output to the different types of display device being employed. The second layer is for "interface control management", this layer provides for terminal device independence in the dialogue. It produces a transparent transformation of the information from the input and output devices into valid dialogue entities to be processed by other layers. The third layer is "dialogue generation", here the elements that constitute the dialogue are constructed. The fourth layer is the "dialogue execution" layer, this is the layer which deals directly with the application and is responsible for overall dialogue management.

Kuo and Konsynski's architecture is a general structure employing knowledge-based techniques for use in any domain of application. Domain dependent knowledge is not directly specified, it is implied in syntactic and semantic knowledge processing identified in the "dialogue execution" layer. The seven knowledge domains identified by Rissland would similarly be diffusely spread within the architecture. However, the primary system for statistical consultation with the user would function as an independent section of the application. The statistical expert system would make use of the interface architecture to communicate with the user, rather than be integrated within it.

INTERFACE STYLE

Conventional MMI (Man-Machine Interface) for statistical software employs a variety of styles. However, the larger packages tend to be dominated by the use of command languages, such as those employed by SAS or SPSS. With this style of interaction the user writes short programs which control the statistical facilities. For example a typical SAS session will consist of two stages, a "data step" and a "procedure step". The "data step" constructs the data-set upon which the "procedure step" performs a statistical procedure. [15] The two steps each consist of a series of program statements. Once the steps have been written they are submitted, either in batch or interactively, to be compiled and executed.

While statistical packages may employ a similar interface design philosophy, and provide access to similar statistical facilities, there is little uniformity in the presentation of the software to the user. This can cause delay and difficulty for users moving to an unfamiliar package to carry out an analysis. The new command names, the scope of their operation, and the logical structures used by the command language, need to be mastered each time a new package is encountered before it can be used for an analysis.

The basic philosophy which supports the conventional statistical interface, is that it provides a means by which the user commands the system to perform specific functions. The intelligent interface on the other hand provides the facilities of an expert assistant. Thus, the style of the dialogue provided could take the form of a co-operative conversational interaction between the interface and its user. [16] Essentially, for a statistical system, the interface and the user would be co-operating to achieve a common goal which would be to obtain the fruits of a statistical analysis. In this relationship the responsibility and final selection of a course of action would rest with the user who would be dominant over the assistant. The assistant would be in a position to criticise the users choice of action and provide an explanation of the reasoning which led to the disagreement with the user. Worden et al [16] have defined the role for the expert assistant as:- "The assistant's job is to point out those things it knows well and which are relevant, to accept the user's overides where he makes them, to work out their consequences, and to try to understand them so it can behave sensibly in future"

CONCLUSION

The most important point which can be made about a useful expert system is not what the system knows, but what its user does not. It is only this difference that makes the expert system of any practical value. Thus it can be seen that no matter how simple or complex an expert system is of itself, it is only of value if it has something to offer the client, no matter how simple or mundane that offering may be within the knowledge domain in which the system operates. An intelligent system for practical problem solving should contain knowledge the client population needs for the task in which it will be used. Thus, it is important to consider the knowledge requirements of the context in which the packages are used, as well as the encoding of the specialist knowledge of the expert into the system.

Intelligent interface design is not simply a question of bolting on an expert system for the domain in which a software package is intended to operate. The scope of the application of knowledge-based techniques can be widened to address identifiable issues in human interface design. Within the domain of statistical technology it is important that the practical constraints faced by the consumer are catered for within the design of the tools he or she is given. In the application of statistics to real world problems the practical difficulties encountered will interact with statistical knowledge to influence the quality of the final analysis.

There is an obvious need for academic research in the area of intelligent statistical interface design, to more clearly resolve the issues of adequately capturing statistical knowledge within an expert system. However, there is also a need for strategic research to more adequately define the requirements of the consumers of statistical technology, for knowledge relevant to the application of the tools with which they are being provided.

REFERENCES

[1] Chambers, J. M. (1981) Some Thoughts On Expert software, Proc. 13th Symposium On The Interface of Computer Science and Statistics, New York, Springer Verlag, 36-40

[2] Hand, D. J. (1984) Statistical Expert Systems: Design, The Statistician, 33, 351-369

[3] Gale, W. A. & Pregibon (1985) Artificial Intelligence Research In Statistics, The AI Magazine

[4] Jamison, W. & Metzler, D. (1985) An Expert System For Statistical Consulting, Proc. 48th American Society For Information Science Annual Meeting, Vol 22

[5] Leeuwenberg, J. (1979) The Cypriots In Haringey, Polytechnic Of North London School Of Librarianship Research Report 1

[6] Furner, S. M. (1983) Response delay time on interactive viewdata information systems, Proc. 10th International Symposium on Human Factors in Telecommunications, Posts and Telecommunications of Finland, 73-79

[7] Kidd, A. L. (1981) Man-Machine Dialogue Design, British Telecom Research Laboratories, Research Study No.1

[8] IBM (1983) Interactive Systems Productivity Facility-Dialogue Management Services-MVS,VM,VSE. Manual SC34 2088 1

[9] Edmonds, E., Guest, S. & Pollard, A. (1984) User Guide For SYNICS/DDL ON VAX Under VMS S.A.I.T. Installation Version 1. Human-Computer Interface Research Unit, Leicester Polytechnic.

[10] Furner, S. M. (1985) The SYNICS Approach To Dialogue Management, The Ergonomist, 182, June

[11] Fink, P. K., Sigmon, A. H. & Berimann, A. W. (1985) Computer Control Via Limited Natural Language, IEE Transactions On Systems Man And Cybernetics

[12] Rissland, E. L. (1984) Ingredients Of Intelligent User Interfaces, Int. J.Man-Machine Studies, 21, 377-388

[13] Deken, J. (1983) Machines And Metaphors, Computer Science And Statistics: The Interface (ed) J. E. Gentle, North-Holland

[14] Kuo, B. & Konsynski, B. (1985) An Architecture For Dialogue Management: Implications in User-Computer Dialogue Design, Interfaces In Computing, 3, 259-275

[15] Cody, R. P. & Smith, J. K. (1985) <u>Applied Statistics And The SAS</u> <u>Programming Language</u>, North-Holland

[16] Worden, R. P., Foote, M. H., Knight, J. A. & Anderson, S. K. (1986) Co-operative expert systems, Report of ESPRIT project P559

Acknowledgement is made to the Director of Research of British Telecom for permission to publish this paper.

PART B
Integrating AI to Stochastic Modelling

6 A Computer Model with Expert Rules for the Control of African Cattle Diseases

G. Gettinby
Department of Mathematics, University of Strathclyde, Glasgow

1   INTRODUCTION

Over the last century the control and eradication of cattle diseases has been one of the major approaches to solving the problems of food production in Africa. Often the solution has been radical. In the early 1900s massive slaughter campaigns involving tens of thousands of cattle were carried out in Southern Africa in an attempt to eradicate rinderpest. More recently technological solutions have been the basis of national and international control programmes. The use of vaccines has averted epidemics of rinderpest, bovine pleuropneumonia and foot and mouth disease. Chemotherapy has controlled infections with other organisms such as rickettsiae.

Initiatives by international aid organisations exist to finance the search for solutions to other major cattle diseases and in particular trypanosomiasis and theileriosis. Trypanosomiasis is a disease not only of cattle but also sheep, goats, camels and wild animals. It is caused by trypanosomes which are transmitted to and from animals by feeding tsetse flies. Theileriosis is also transmitted by an intermediate vector. it is primarily a disease of cattle which is widespread in countries on the East coast of central Africa where it is more commonly known as East Coast fever (ECF). The disease is caused by a protozoan parasite (*Theileria parva*) which is transmitted between cattle by ticks (*Rhipicephalus appendiculatus*) which feed on the blood of the cattle.

Most parasite life-cycles are complex. In the case of ECF and trypanosomiasis the life-cycles of the intermediate tick and fly vectors, and the infectious parasites they transmit are still not properly understood. Basic research programmes are in progress by different institutes to find immunological solutions which will enable cattle to be vaccinated against the vectors or the disease, and to find prophylactic drugs which will prevent cattle from becoming infected. As specialised knowledge is acquired, and the implementation of any control programme in the field is contemplated, the unification of the accumulated knowledge of expert immunologists, veterinary pathologists, vector ecologists, economists, epidemiologists etc. becomes essential to obtain an overall perspective. Otherwise decisions are based on dogma or incorrect perceptions outwith the knowledge of any single expert. Unfortunately no such structure for the pooling of expertise exists. This paper describes the development of a computerised system ECFXPERT for the study of ECF based on the mathematical relationship between observations made by past and present experts.   It is intended to serve as a test model for the integration of expert knowledge and the regulation of scientific inquiry into the control of ECF and other parasitic diseases by pesticides, vaccination and chemotherapy. Using ECFXPERT, experts in any relevant area can use computer experiments to test the consequences of their observations and judgements in a system which contains complementary expert knowledge from other disciplines.

2   MATERIALS AND METHOD

East Coast Fever

The life-cycle of the tick consists of four successive stages: egg, larva, nymph and adult. The larvae emerge from the eggs and attach to animal hosts in order to feed. After engorgement the larvae drop to the ground and undergo development to the nymph stage. In a similar fashion the nymph finds an animal host on which to feed, and develops to the adult stage. Before reproduction the male and female adult ticks must attach and feed, and thereafter the female returns to the herbage where she lays eggs before dying. It is during the larval and nymphal feeding bouts that the tick can become infected. This infection is then passed on to other animals during feeding in the next stage. The parasite leaves the gut of the tick and enters the bloodstream of the cattle where it replicates inside the white blood cells. Eventually the parasite infects the red blood cells in a form known as a piroplasm. Ticks feeding on the blood cells of cattle become infected with these piroplasms. Several weeks after infection untreated cattle normally die. Those cattle which survive may have immunity against future infections.

The main method of control has been to dip cattle in an acaracide to kill attached ticks. Recent research has demonstrated that in addition to successful drug chemotherapy a method of vaccination known as 'infect and treat', first contemplated over 50 years ago, offers the prospect of widespread immunization of all types of cattle against the diseases (Irvine, 1985). Several international aid programmes are currently funding extensive field trials in this area.

Complications do arise. Cattle immune to local strains of ECF are not necessarily immune to all strains. Immune animals can be carriers of the infective piroplasms and so transmit the disease. In village and ranch management systems cattle and wild buffalo often graze the same pasture. Wild buffalo carry an infection similar to ECF, which is transmitted by ticks in the same way. The ticks can cross-infect cattle with the buffalo disease from which the cattle usually die. The full model has been designed to take account of these factors but for purposes of illustration these complications are not dealt with in the model described herein.

ECFXPERT

ECFXPERT consists of three parts. The <u>primary knowledge base</u> contains an inventory specifying the rules for disease transmission and on which experts are in agreement. As a complement to the expert rule inventory the primary knowledge base also contains suggested default assignations. These are the set of confidences and parameters which the user may assign in the absence of an opinion. They have been inferred from historical expert knowledge as reported in scientific journals of from opinions expressed by current experts working in the area.

The <u>secondary knowledge base</u> consists of the entries made by the user to the questions arising from the rules in the primary knowledge base. This information is acquired from the user during computer interrogation.

The structured knowledge base consists of a set of mathematical models which manipulate the knowledge in the primary and secondary knowledge bases to provide inferences from which decisions can be made.

Expert rule inventory

Tick life-cycle.   The life-cycle of the tick consists of the following parts - egg, unattached larva, attached larva, engorged larva to nymph, unattached nymph, attached nymph, engorged nymph to adult, unattached adult male, unattached adult female, attached adult male, attached adult female, preoviposition.

Duration of the non-development parts of the tick life-cycle.   There is considerable agreement among experts on the reported findings for these durations:

> unattached larva - the minimum duration is 13 days, thereafter the larva dies or attaches with a preassigned confidence (Branagan, 1973a),

> attached larva - the duration is 5 days (Branagan, 1973a),

> unattached nymph - the minimum duration is 17 days, thereafter the nymph dies or attaches with a preassigned confidence (Branagan, 1973a),

> attached nymph - the duration is 6 days (Branagan, 1973a),

> unattached adult - the minimum duration is 21 days, thereafter the adult dies or attaches with a preassigned confidence (Branagan, 1973a),

> feeding adult female - the duration is 9 days (Newson, per comm),

> feeding adult male - the duration is 18 days (Newson, per comm).

Tick survival.   Newson, Chiera, Young, Dolan, Cunningham and Radley (1984) report observations on the survival and longevity of larvae, nymphae, and adult ticks under field conditions. From these data daily survival rates have been inferred:

> in the first 175 days of the larval stage the daily survival rate is 0.996 and 0.926 thereafter,

> in the first 270 days of the nymphal stage the daily survival rate is 0.997 and 0.980 thereafter,

> in the first 400 days of the female adult stage the daily survival rate is 0.998 and 0.982 thereafter,

> . in the first 300 days of the male adult stage the daily survival rate is 0.998 and 0.977 thereafter.

Tick fecundity.   The tick population is maintained in a steady-state by assigning a value to the fecundity of a female during oviposition which adjusts for all ticks lost in the previous life-cycle due to natural mortality. The ratio of female to male ticks is 1:1 (Branagan, 1973a).

Tick mating.   Mating takes place on the host and male ticks are polygamous:

> male and female ticks feed for 4 days in order that sexual development can take place before mating (Newson, per comm),

mating occurs within 21 days from attachment otherwise ticks become dehydrated (Newson, per comm),

male ticks can mate several female ticks (Young, Newson, per comm),

the ratio of female to male ticks on cattle is 1:2 (Kaiser, Sutherst and Bourne, 1982).

Animal status.   It is necessary to distinguish between the different stages of the disease in cattle:

an infected cow becomes an infective 10 days after acquiring the infection (Branagan, 1969),

an infected cow either dies or recovers (with a preassigned confidence) 25 days after acquiring the infection (Lewis, 1950; Branagan, 1969; Dolan, Young, Leitch and Stagg, 1984),

an infected cow which recovers becomes immune 42 days after acquiring the infection (Branagan, 1969),

an immune cow may be a carrier (with a preassigned confidence) (Young, Leitch and Newson, 1981).

Animal/tick disease transmission.

an uninfected larva tick feeding on an infective cow becomes infected (with a preassigned confidence),

an uninfected nymph tick feeding on an infective cow becomes infected (with a preassigned confidence),

an uninfected larva or nymph tick feeding on a carrier becomes infected (with a preassigned confidence),

an infected tick loses its infection when it feeds on an uninfected cow and the cow becomes infected (with a preassigned confidence) (Lewis, 1950; Brocklesby, Barnett and Scott, 1961),

an infected tick feeding on an infective cow loses its infection, acquires a new infection and the cow remains infective,

an infected tick loses its infection and acquires a new infection when it feeds on a carrier (with a preassigned confidence).

Interrogation with default assignations

If the rules are to be used to predict the course of disease in cattle it is necessary to prescribe grazing conditions, tick challenge, meteorological factors and preassigned confidences for the transmission of infection etc.   For this purpose the system provides a questionnaire. Each question is accompanied by a brief explanation and where known a suggested default assignation. Many of the questions accept a confidence response to allow users to exercise their subjective expert judgement.

1. Do you have cattle?

   [ECF is a parasitic disease of cattle]

2. Estimate, to the nearest 10, how many cattle you have:

   [ECF is transmissable between cattle and incidence depends on the number of cattle]

3. Estimate how many cattle are currently infected with ECF:

   [The spread of the disease will depend on the number of infected cattle]

4. Estimate how many cattle have immunity against ECF:

   [Although Piercy (1947) reports an isolated incidence in which an immune animal died from reinfection 2 years later, it is generally believed that immune cattle cannot die from subsequent ECF infections]

5. Do the immune cattle have carrier status?

   ["Carrier status" means that recovered animals remain infective. Wilde (1967) suggests that carrier status is rare. Barnett and Brocklesby (1966) acknowledge its existence. Young, Leitch and Newson (1981) have shown it can occur after natural infection and Dolan (1981) after chemotherapy]

6. Do you have *R. appendiculatus* ticks present where cattle graze?

   [These ticks are necessary for the transmission of ECF to and from cattle. ECF can only occur in areas where these ticks infest the grazing land (Branagan, 1978)]

7.(a) Estimate how many of these ticks infest the area grazed by the animals:

   [Ticks feed on cattle and other animals. This allows transmission of the parasite which causes ECF from one animal to another. The spread of the disease will depend on the number of ticks]

7.(b) Estimate how many of these ticks are

   EGGS   LARVAE   NYMPHS   ADULTS

   [As the ECF infection cannot effect transovarial passage the infection can only be picked up by feeding larvae or nymphs and passed on by feeding nymphs or adults. The breakdown of the initial population is thus important]

7.(c) What percentage of this tick population (0-100) do you believe to be infected with ECF at the present time:

   [This value is generally accepted to be very low. If it is unknown a value of 1 is suggested based on the findings of Barnett (1956)]

8. Enter the date as a day of the year 1-365:

   [The percentage of ticks in each stage will change with time]

9.(a) With what confidence (0-1) do you believe that uninfected larvae acquire infection when feeding on infective animals:

   [If this is unknown then 0.35 is suggested. This value is obtained from the studies of Cowdry and Ham (1932)]

9.(b) With what confidence (0-1) do you believe that uninfected nymphs acquire infection when feeding on infective animals:

   [If this is unknown then 0.6 is suggested. This value is obtained from the studies of Cowdry and Ham (1932)]

10.(a) With what confidence (0-1) do you believe that uninfected larvae acquire infection while feeding on immune animals with carrier status:

10.(b) With what confidence (0-1) do you believe that uninfected nymphs acquire infection while feeding on immune animals with carrier status:

[The infection rate of larval and nymphal ticks feeding on immune animals is unknown. It is suspected that immune animals have fewer circulating piroplasms than infective animals and there will be less chance of a feeding tick picking up infection. If you wish a low infection rate a confidence of 0.01 is suggested. If you wish sterile immunity enter a confidence of 0]

11. With what confidence (0-1) do you believe that an infected tick feeding on an uninfected animal passes on the infection:

[If this is unknown 0.876 is suggested. This value is based on the findings of Barnett and Brocklesby (1966)]

12. With what confidence do you believe that a cow infected with ECF will subsequently die:

[If unknown 0.95 is suggested. This value is based on the findings of Barnett and Brocklesby (1966). All the animals which recover become immune to ECF. The response of an animal to ECF infection is thought to depend on nutrition, breed, sex etc. (Wilde 1967)]

13.(a) With what confidence (0-1) do you believe that larvae which seek hosts on a particular day are successful:

13.(b) With what confidence (0-1) do you believe that nymphs which seek hosts on a particular day are successful:

13.(c) With what confidence (0-1) do you believe that adults which seek hosts on a particular day are successful:

[Tick populations are often localised and so cattle may not always be accessible to questing ticks. If this value is unknown a confidence of 0.1 is suggested]

14. Specify the number of days over which the spread of disease in the cattle and tick populations is to be predicted:

[You can be more confident about the results from short term predictions. Should you require predictions for periods greater than a year, the simulation assumes the same monthly temperatures recur]

15. Enter climate details

The details required are the mean daily maximum and mean daily minimum temperatures for each month of the year, together with the standard deviations of these values

Month	Max	Max s.d.	Min	Min s.d.
Jan				
.				

[Muguga, Nairobi, Kenya 24-year averages are:

Month	Max	Max s.d.	Min	Min s.d.
Jan	22.6	1.52	11.1	1.63
Feb	.	.	.	.

These can be used if thought appropriate for your region]

Mathematical models

Development model.   Studies on the tick have demonstrated that the development parts of the life-cycle are regulated by climate with warm humid conditions being most favourable (Browning, 1976). Branagan (1973b) has shown that the rate of development increases with temperature in a predictable manner, whereas relative humidities of less than 40% may marginally retard development.

The predictions of the duration of the development parts under field conditions can be obtained using the weighted development fraction technique.   This approach has been successfully demonstrated for other parasite systems (Gettinby, Bairden, Armour and Benitez-Usher, 1979; Gardiner, Gettinby and Gray, 1981). Branagan (1973b) reports on the durations of the development parts: preoviposition, egg to larva, engorged larva to nymph, and engorged nymph to adult, at each of the constant temperatures 15, 18, 21, 25 and 29 C. From these observations the relationship between duration d(T) and constant temperature T for each development part has been established using least-squares methods. The relationships are shown in Table 1.

TABLE 1   Summary of development functions relating constant temperature T and development time d(T) for different development parts of the tick life-cycle.

Phase of life cycle	Temperature $^{\circ}$C	Form of d(T)
Pre-oviposition	$T < 15$	$\exp(1/(0.00149 + 0.0193T))$
	$15 \leq T < 18$	$121 - 6T$
	$18 \leq T < 21$	$37 - 1.33T$
	$21 \leq T < 29$	$24.75 - 0.75T$
Egg to larva	$T < 15$	$1/(0.00109T - 0.00738)$
	$15 \leq T < 18$	$280 - 11.33T$
	$18 \leq T < 21$	$220 - 8T$
	$21 \leq T < 29$	$115 - 3T$
Larva to nymph	$T < 15$	$1/(0.00109T - 0.00738)$
	$15 \leq T < 18$	$1/(0.00258T - 0.01786)$
	$18 \leq T < 21$	$1/(0.00899T - 0.13333)$
	$21 \leq T < 25$	$47.5 - 1.5T$
	$25 \leq T < 29$	$37 - T$
Nymph to adult	$T < 15$	$1/(0.00139T - 0.0075)$
	$15 \leq T < 18$	$1/(0.00111T - 0.00333)$
	$18 \leq T < 21$	$1/(0.00555T - 0.08333)$
	$21 \leq T < 25$	$93.3 - 3T$
	$25 \leq T < 29$	$49.25 - 1.25T$

Typically daily temperature starts at a minimum, T1, rises to a maximum, T2, at midday and falls to a second minimum, T3, at the end of the day. Denoting the start, midpoint and end of the day by t = 0, 1/2 and 1 respectively, the daily development fraction is calculated using

$$F = \int_0^{1/2} \frac{1}{d(T)}\, dt + \int_{1/2}^{1} \frac{1}{d(T)}\, dt$$

The relationship between temperature T and time of day t is given by the sinusoidal function

$$T = A1 + B1 \sin(2\pi t - \pi/2)$$

where
$$A1 = \begin{cases} \dfrac{T1+T2}{2} & 0 \leq t < 1/2 \\[2ex] \dfrac{T1+T3}{2} & 1/2 \leq t \leq 1 \end{cases}$$

and
$$B1 = \begin{cases} \dfrac{T2-T1}{2} & 0 \leq t < 1/2 \\[2ex] \dfrac{T2-T3}{2} & 1/2 \leq t \leq 1 \end{cases}$$

The evaluation of each of the above daily development fractions is by numerical integration. Summing the development fractions associated with each day experienced by the tick during a development part, the development is predicted complete on the first day when the sum exceeds 1.

The prediction rule takes the form:

> if the sum of the daily development fractions is first greater than 1 after n days

> then the development part is predicted complete after n days

The influence of relative humidity on development is considered not to be limiting (Branagan, 1973b; Newson, 1978).

Attachment model. In order to feed ticks must find a host. Successful attachment to cattle depends on the stocking rate. Unattached questing ticks find an animal to feed upon with a preassigned confidence. If this confidence is unknown it is assumed that the probability of a tick attaching in time dt is $\alpha r\, dt$ where r is the cattle stocking rate per tick. P(t) the probability of a tick attaching in time t is then given by:

$$P(t+dt) = P(t) + (1-P(t))\alpha r\, dt.$$

If P denotes the probability of a tick attaching on any day the above equation leads to

$$P = 1 - \exp(-\alpha r).$$

The constant α will depend on whether the questing tick is a larva, nymph or adult.

Temperature simulation model. Using the monthly means and standard deviations entered by the user or the default assignations, typical maximum and minimum temperatures for each day of the year are simulated using random numbers generated from a fitted extreme value distribution (Morgan, 1985).

Tick life-cycle and disease simulation model. Commencing with the conditions prescribed by the user during interrogation, daily changes in the tick and cattle populations are predicted. A record is kept of the progression of each tick through its life-cycle. For binary events which occur with a preassigned confidence such as on a particular day whether or not a questing tick attaches or the tick survives, the outcome of the event is determined at random using random numbers from the Bernoulli distribution. A record is kept of the disease status of each animal and tick. The transmission of disease to and from cattle and ticks with preassigned confidence is similarly treated as a binary event and the outcome determined by lottery using the Bernoulli random number generator. The use of random number generators to simulate random events produces a close representation of the variation experienced by ticks, cattle and parasites under field conditions.

Programming and inferences

The current system has been developed for use on an IBM XT and uses Prospero Pascal. The default assignation text is not compiled in the simulation programme but maintained separately as a text file which is accessed at run time. This text read facility allows experts to revise and update the default assignations using a word processor.

The system provides a graphics representation of the spread of the disease in the cattle population. This representation illustrates the day to day changes in the distribution of uninfected, infected, infective and immune animals. In addition the number of uninfected and infected ticks are listed, and the number of ticks in each stage of the life-cycle.

Summary statistics of animal mortality, morbidity and immunity are listed during and at the end of the prediction period. Currently the user can choose from a menu offering:

  (i) repeated control runs;

 (ii) an investigation into the effect of changing some, or all, of the confidences associated with factors, a list of factors in order of influence on morbidity is produced;

(iii) an investigation into the influence of one factor over a range of confidences, a table of confidence factor as a function of morbidity is produced;

 (iv) computer experiments of control against acaracidal treatment to test the hypothesis that acaracide has no effect, confidence intervals for incidence of morbidity and a test of significance are produced.

3   RESULTS OF AN APPLICATION

The following example illustrates some of the statistics generated by ECFXPERT.

A herd of 100 cattle are grazed on pasture infested with 1000 ticks of which 1% are initially infected. Grazing commences at the beginning of March (day 60) with 10 infective cattle and predictions are made over a 100 days using the default assignations. The meteorological data used are 24-year monthly averages of daily maximum and minimum temperatures at Muguga, Nairobi. Using ECFXPERT the mortality, morbidity and immunity rates were estimated to be 23%, 13% and 1% respectively.

For an arbitrary 5% increase in all confidences, using ECFXPERT the factor found to be most influential in increasing morbidity was the percentage of ticks initially infected, whereas the factors relating to the infection rate of ticks from immune cattle, infection rate of animals from infected ticks and the tick attachment rates were least influential.

An examination of the most influential factor, the percentage of ticks initially infected, over the confidence range 0 to 10% produced the following results:

% infected	0	2	4	6	8	10
% morbidity	6	37	38	46	62	71

This suggests that even a low proportion of initially infected ticks can produce a high incidence of disease.

Computer simulations of five repeated control and five repeated acaracidal experiments led to a 95% confidence interval 15% $\pm$ 5% for the mean rate of morbidity from the control experiments and 2.2% $\pm$ 2% from the acaracidal experiments. Acaracidal dipping was applied at 14 day intervals and assumed to be 90% effective in killing ticks up to 8 days after application.

4   DEVELOPMENTS

ECFXPERT demonstrates that a computer model for East Coast fever is a viable concept. A great deal more work needs to be undertaken to calibrate the system using field observations. Expert rules from drug chemotherapy and vaccination research programmes need to be established and models for the dynamics of the parasite within the gut of the tick and the bloodstream of the cow need to be developed. Further consideration of the programming environment is necessary to enable the rule structure to be checked for inconsistencies and rules to be added without the labour of difficult programming.

The application of computer models with expert rules for the control of infectious diseases of animals is immense. Tick-transmitted diseases are widespread throughout the world. In Africa, heartwater, anaplasmosis, babesiosis and some viral diseases (e.g. Nairobi sheep disease) are all diseases similarly transmitted by ticks. Tick pyaemia, babesiosis, louping ill and tick-borne fever are endemic diseases of cattle and sheep in Europe which are spread by the three-host tick *Ixodes ricinus* in a fashion similar to that of East Coast fever. In Australia the one-host tick *Boophilus microplus* is a major vector of disease. Most of these diseases could be investigated using the ECFXPERT shell. Other prospects for the control of infectious diseases such as trypanosomiasis (Murray and Gray, 1984), which involve a fly vector, could also be studied.

5   ACKNOWLEDGEMENTS

The author is indebted to the advice and enthusiasm given by many experts during consultations and in particular: Dr. R. Newson of the International Centre for Insect Physiology and Ecology, Nairobi; Dr. A.D. Irvine, Dr. S. Morzaria, Dr. T.T. Dolan and Dr. J. Doyle of the International Laboratory for Research on Animal Diseases, Nairobi; Dr. A.S. Young of the Kenya Agricultural Research Institute, Muguga, Kenya. D. King is acknowledged for his computational assistance and M.V. Thrusfield for constructive criticism. The author wishes to thank the Director General of the International Laboratory for Research on Animal Diseases, Nairobi for the opportunity to visit and work at ILRAD where much of this work was undertaken.

REFERENCES

1. Barnett, S.F. (1956). Annual report of the East African Veterinary Research Organisation, 1955-56. Government printer, Nairobi.

2. Barnett, S.F. and Brocklesby, D.W. (1966). A mild form of East Coast fever (*Theileria parva*) with persistence of infection. Br. Vet. Jour. 122, 361-370.

3. Branagan, D. (1969). The maintenance of *Theileria parva* infections by means of the Ixodid tick *Rhipicephalus appendiculatus*. Trop. Anim. Hlth. Prod. 1, 119-130.

4. Branagan, D. (1973a). Observations on the development and survival of the Ixodid tick *Rhipicephalus appendiculatus* (Neumann, 1901) under quasi-natural conditions in Kenya. Trop. Anim. Hlth. Prod. 5, 153-165.

5. Branagan, D. (1973b). The development periods of the Ixodid tick *Rhipicephalus appendiculatus* (Neumann) under laboratory conditions. Bull. Entemol. Res. 63, 155-168.

6. Branagan, D. (1978). Climate and East Coast Fever. In: Weather and Parasitic Animal Disease. pp. 126-140.

7. Brocklesby, D.W., Barnett, S.F. and Scott, G.R. (1961). Morbidity and mortality rates in East Coast Fever (*Theileria parva* infections) and their application to drug screening procedures. Br. Vet. Jour. 117, 529-531.

8. Browning, T.O. (1976). The aggregation of questing ticks, *Rhipicephalus pulchellus* on grass stems with observations on *Rhipicephalus appendiculatus*. Physiol. Ent. 1, 107-114.

9. Cowdrey, E.V. and Ham, A.W. (1932). Studies on East Coast Fever. 1. The life-cycle of the parasite in ticks. Parasitology, 24, 1-49.

10. Dolan, T.T. (1981). Progress in the chemotherapy of theileriosis. In Advances in the Control of Theileriosis. A.D. Irvin, M.P. Cunningham and A.S. Young, Eds., Martinus Nijhoff Publishers, The Hague, 186-208.

11. Dolan, T.T., Young, A.S., Leitch, B.L. and Stagg, D.A. (1984). Chemotherapy of East Coast fever: Parvaquone treatment of clinical disease induced by isolates of *Theileria parva*. Vet. Parasitol. 15, 103-116.

12. Gardiner, W.P., Gettinby, G. and Gray, J.S. (1981). Models based on weather for the development phases of the sheep tick, *Ixodes ricinus* L. Vet. Parasitol. 9, 75-86.

13. Gettinby, G., Bairden, K., Armour, J. and Benitez-Usher, C. (1979). A prediction model for bovine ostertagiasis. Vet. Rec. 105, 57-59.

14. Irvin, A.D. (1985). Immunity in Theileriosis. Parasitology Today, 5, 124-128.

15. Kaiser, M.N., Sutherst, R.W. and Bourne, A.S. (1982). Relationship between ticks and Zebu cattle in Southern Uganda. Trop. Anim. Hlth.

Prod. 14, 63-74.

16. Lewis, E.A. (1950). Conditions affecting the East Coast Fever parasite in ticks and cattle. E. Afr. Agric. J., 16, 65-77.

17. Morgan, B.J. (1985). Elements of Simulation. Chapman and Hall.

18. Murray, Max and Gray, A.R. (1984). The current situation on Animal Trypanosomiasis in Africa. Preventive Veterinary Medicine, 2, 23-30.

19. Newson, R.M. (1978). The life-cycle of *Rhipicephalus appendiculatus* on the Kenyan coast. In: Tick-borne diseases and their vectors. Ed. J.K.H. Wilde. pp.46-50.

20. Newson, R.M., Chiera, J.W., Young, A.S., Dolan, T.T. Cunningham, M.P. and Radley, D.E. (1984). Survival of *Rhipicephalus appendiculatus* (Acarina: Ixodidae) and persistance of *Theileria parva* (Apicomplexa: Theileriidae) in the field. Inter. J. Parasitology, 5, 483-489.

21. Piercy, S.E. (1947). Immunity to East Coast Fever. The Veterinary Record, Vol.59, No.46, 636.

22. Wilde, J.K.H. (1967). East Coast Fever. Advances in Veterinary Science. 11, 207-259.

23. Young, A.S., Leitch, B.L. and Newson, R.M. (1981). The occurrence of a *Theileria parva* carrier state in cattle from an East Coast fever endemic area of Kenya. In Advances in the Control of Theileriosis. A.D. Irvin, M.P. Cunningham and A.S. Young, Eds., Martinus Nijhoff Publishers, The Hague, 60-62.

7 AI and Stochastic Process Simulation

Ray J. Paul
London School of Economics

1. INTRODUCTION

Simulation is a set of techniques for using a computer to imitate the operations of various kinds of real-world facilities or processes. In order to study scientifically the facility or process of interest, we often have to make a set of assumptions about how it works. These assumptions, which usually take the form of mathematical or logical relationships, constitute a model which is used to try and gain some understanding of how the corresponding system behaves.

Simulation is used to study models of real-world problems that are too complex to be evaluated analytically. In a simulation we use a computer to evaluate a model numerically over a time period of interest to estimate the desired true characteristics of the model. For example, we have simulated a port which handles a variety of incoming and outgoing cargoes using a mixture of specialised and general berths in order to aid the planning of future berth requirements [8]. The simulation model was used to simulate the operation of the port as it currently exists and as it would be if the port were expanded or not.

Simulation then, involves the setting up of a model of the system under study, in which all the relevant components are defined, and the way in which they change through time and effect each other are exactly specified. Typically the simulation model is stochastic because it will contain several random variables (for example, the inter-arrival times of ships at a port may be random). The model is set in motion and its behaviour observed. The output data for a stochastic process simulation model are themselves random and thus are only estimates of the true characteristics of the model. By running the model for a set time, the values taken by the output variables can be compared to the values taken by the corresponding variables in the real system. If the correspondence is close, then the model may be considered to be a good representation of reality. The model then provides a potentially powerful tool for conducting controlled experiments, by systematically varying specific parameters and re-running the model.

Simulation is an important modelling technique as evidenced by the survey conducted by Beasley and Whitchurch [3]. However, simulation is an expensive technique to use in spite of recent advances in computing hardware and software. The survey by Christy and Watson [5] of industrial practice concerning simulation application, shows that cost as expressed in terms of software is still one of the major disadvantages of using simulation modelling. What is clear from both papers is that simulation modelling is a powerful method for modelling problems and it would be more widely used if it became cheaper and easier to do so.

Simulation modelling has always needed and employed a high level of computer support. The practice of simulation has grown parallel with the development of computing power. Simulation modellers were quick to recognise the benefits of specialised simulation structures, often embedded in special-purpose, high-level languages in promoting the faster construction and testing of large models. More recently program generators have further speeded up the process of model coding. Microcomputers have allowed the modeller literally to take the model rather than simply a pile of computer output to the user for verification. Computer graphics have also been used to improve the end-user acceptability of simulation models and their results.

However, simulation remains an expensive technique. Greater processing power has encouraged the wider use of simulation rather than possible analytic alternatives. More ambitious models encompass ever greater volumes of detail. The relative expense of simulation is now largely a function of analyst rather than computer time and the pressure to improve the productivity of the analyst increases and will continue to do so.

The Computer Aided Simulation Modelling (CASM) project at the London School of Economics (LSE) has been set up to research into ways of automating parts of the process of simulation modelling. An outline of the CASM research is given in Balmer and Paul [1]. In this paper, we shall look at the research into artificial intelligence techniques to aid simulation modelling. First we give an overview of CASM research. Three aspects of artificial intelligence in stochastic process simulation are concentrated on: automatic program generation, model formulation, and output design and analysis.

2.   COMPUTER AIDED SIMULATION MODELLING

The CASM project team has been working upon a flexible plan for developing computer aids to simulation modelling. Here we develop an overview of this plan.

Figure 1 illustrates the basic process of simulation model development as envisaged by CASM. The analyst formulates the problem in some structured way, for example as an activity cycle diagram (ACD) or a flowchart. ACDs are described by Clementson [6] and Poole and Szymankiewicz [19] among others. The model logic thus defined is fed into an interactive

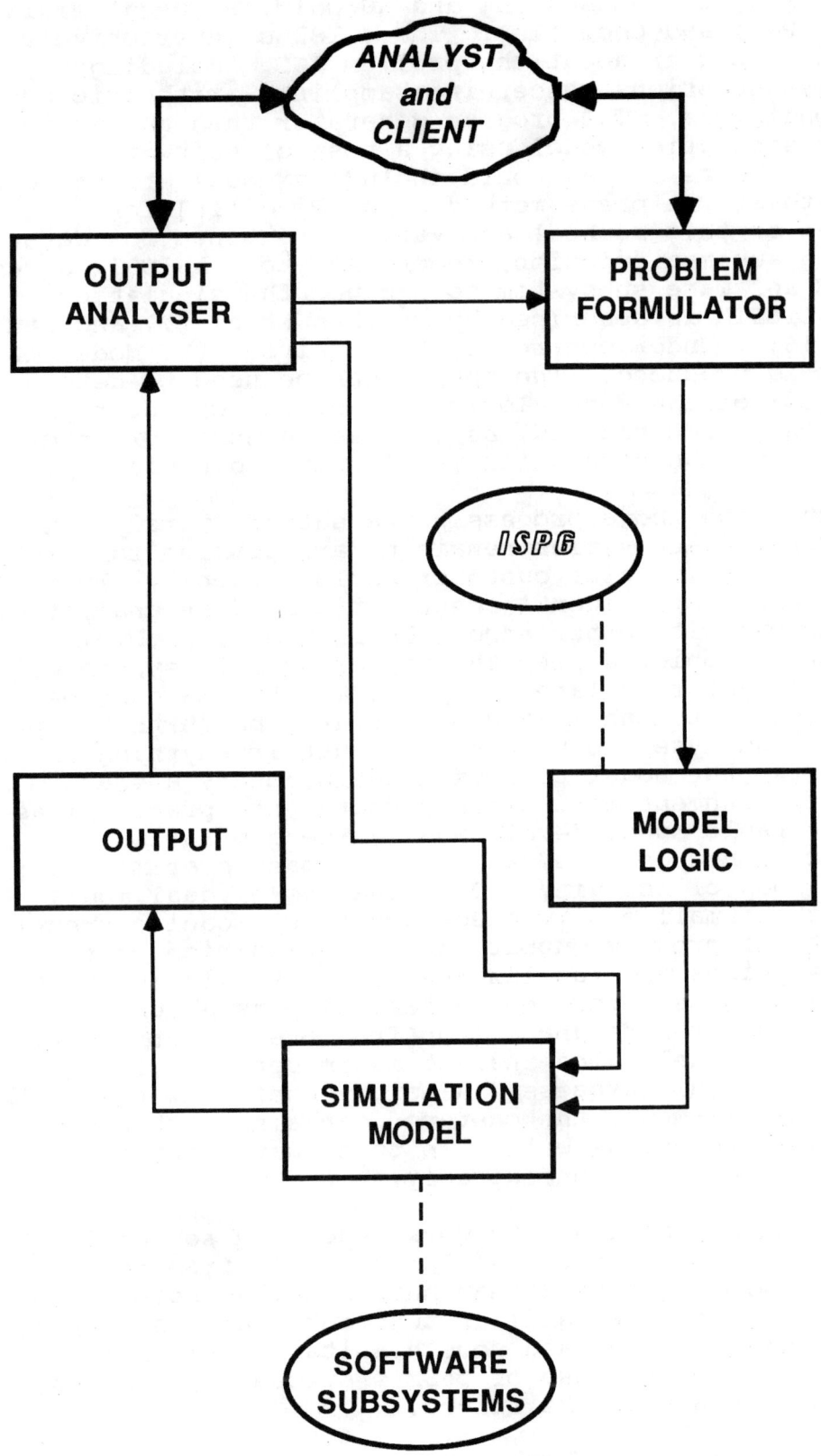

Fig. 1.   The process of simulation modelling

simulation program generator (ISPG). Several ISPGs exist, such as CAPS by Clementson [6] and AUTOSIM, a Pascal emulation of CAPS by Paul and Chew [16]. These ISPGs interactively interrogate the user about the problem ACD, including quantitative questions concerning sampling, arithmetic and initial conditions. The program generator then automatically writes the simulation model using a host of software subsystems. These latter would include a model structure (in our case, the three phase method - see Pidd [18] for a description of this method) and various routines for data sampling, queue manipulation, recording etc. AUTOSIM accesses the LIBSIM software subsystem to produce the simulation program. LIBSIM is described by Crookes et al [7] and Paul and Chew [16]. Under control of the analyst, the model is run and output is produced. The output can be used to determine 'correctness' of the model logic, and of the computer program (if the ISPG is not trusted) as well as reruns. Graphics is used to emulate the simulation model output dynamically.

Assuming that the above process works satisfactorily, the labour intensive activities remaining are problem formulation and output analysis (plus customer satisfaction!). These activities are 'intelligent' contributions of an analyst which tend to improve with experience. In figure 1, problem formulation is depicted with the aid of an A.I. system which helps the analyst formulate the problem with the customer. The expertise of the analyst can similarly be further captured in an output analyser to help decide what if anything is wrong and how to run the model to obtain satisfactory answers. The modelling environment more closely represents practical as well as desirable model development. The simulation environment is not depicted as a single pass system, but as a continuous loop of activity. This enables gradual model development in small easily checked stages, model correction in the light of program output, and determination of the running conditions and run lengths of the simulation model. The latter could be determined dynamically as a function of output and hence the feedback loop from the output analyser to the simulation model. The analyst is in control of and participates in this process. A major benefit is that with fast model development, the customer can also participate in the modelling process as well. In these ways, integration of the system is advanced considerably.

One of the dangers of the CASM work described so far is the ease with which research can distance itself from reality. With this in mind, the CASM team have modelled several real problems as well as the familiar textbook examples. Apart from some defence models and some hospital models, El Sheikh et al [8] report on the use of CASM's systems for analysing port traffic now and in the future.

Work on the problem formulator has been in progress from an early date, and a first attempt at using an expert system was described by Doukidis and Paul [10]. Expert systems turned out to be inappropriate for handling what was essentially a natural language understanding problem, and a recent paper by Paul and Doukidis [17] updates the previous publication. CASM

is now researching into a Natural Language Understanding
System (NLUS) and progress is good. The research that led
from an expert system to a N.L.U.S. is described by Doukidis
and Paul [11].

3.   AUTOMATIC PROGRAM GENERATION

Barr and Feigenbaum [2] define automatic programming as the
automation of some part of the programming process. Some
success has been achieved in this area with systems that help
programmers manage large programs or that produce programs
from some specification of what they are to do (besides the
code itself). AUTOSIM is an example of the latter type of
application.

AUTOSIM is an ISPG which accepts a model specification in
terms of an ACD representation and produces a simulation
program written in Pascal and supported by LIBSIM. It is
written in VAXPascal on a VAX machine and in Turbo Pascal on
an IBM microcomputer.

The AUTOSIM system reflects the general structure of an ISPG
in its two distinctive components, namely the interactive
input of a model specification and the generation of a Pascal
simulation program. Corresponding to these two components
AUTOSIM produces two output files. First there is the
interactive data file, which consists of all the input data
which is extracted with the help of an ACD through an
interactive input session. Second, there is the generated
Pascal program file, which contains the Pascal simulation
program generated by using as input the interactive data file.

The interactive input session is intended to capture the logic
of the proposed simulation model. The description of the ACD
and associated data completing the model specification is
elicited via an interactive dialogue with the computer. The
responses of the user are used to guide the subsequent
questionning, to provide a simple analysis of the system
described as well as to assemble a formal model specification.

The system requires the specification of the names of each
type of object or entity in the model. The life cycle of each
entity is then defined by a sequence of alternating active and
passive states called activities and queues respectively.
Following the entry of the cycles the system permits a review
of the specification so far and gives some simple analysis of
the model logic.

The next part of the session allows the definition of
particular queueing disciplines for the various queues defined
during the description of the life cycles. Where pertinent
the relative priority of the activities mentioned in cycle
descriptions can be defined. The next series of questions
relate to the durations of activities (typically sampled from
a statistical distribution) and the arithmetic of any entity
attributes. The data collection requirement of the simulation
model is defined in terms of the production of histograms of
attribute values or of queue lengths or of queue waiting times.

Finally the starting conditions required are specified. The
initial disposition of entities within their life cycles is
defined. The times of completion of any activities deemed to
be in progress at time zero are also requested.

These data defining the model specification are then stored in
a named text file for subsequent use. The generation of a
simulation program from a previously defined model
specification file requires the user merely to select the
appropriate option and the relevant data file name. The
program can be written in Vax Pascal or Turbo Pascal and will
use the same LIBSIM routines in each case.

AUTOSIM has been tested on a number of real life problems and
has been used by students at the LSE in simulation project
work. As was previously mentioned, it is not always possible
to specify precisely the desired model within the interactive
format provided by AUTOSIM and so the program code generated
does not always match exactly the original problem. However,
the program, being so well-structured, is easily modified (not
all students believe this!) to accommodate those features
inaccurately represented. Although some handcrafting has
proved necessary the larger part of the coding task is
performed by AUTOSIM. A full description of how AUTOSIM works
is given by Chew [4] and Paul and Chew [16].

4.   PROBLEM FORMULATION

In this section we look at the N.L.U.S. that was developed as
an aid to simulation problem formulation. Paul and Doukidis
[17] propose a classification scheme for N.L.U.S.s based on
their domains, their dictionaries, and their independence of
other systems. The simulation N.L.U.S. operates within the
family domain of defining problems for simulation analysis.
That is, a large class of problem can be addressed by the
system since such problems have many similarities. Although
the sorts of problems we are concerned with are largely
queueing problems, such problems arise in many walks of life
such as hospitals, banks, ports, assembly lines etc. However,
full knowledge of the problem cannot be anticipated as could
be for a system designed for a specific domain. Hence the
dialogue of the system is restricted.

The dictionary of the simulation N.L.U.S. typifies the system
as a learning system. By this we mean that the dictionary
starts small, is frequently updated, and has structure. The
initial dictionary is specific to simulation modelling and is
expanded as the problem tackled is processed. Sometimes
different verbs which represent the same action have to be
grouped together. Hence the dictionary is flexible and allows
the manipulation of these specific cases.

The third part of our classification is independence. In our
case, the system is a sub-system of other systems used in the
simulation modelling process. Output from the N.L.U.S. can be
used with AUTOSIM to produce a working simulation program
representing the original problem. Although work on a
graphics enhancement is in hand, it is unrealistic to expect

the system ever to be useful in any stand-alone mode.

The N.L.U.S. accepts an informal specification of the problem
in a form of English sentences. Because of the nature of such
problems, the system accepts a specific set of sentences
called "action sentences". For example:
>   The patient is treated by the doctor.
>   The ship unloads at a berth.
>   The mechanic disconnects a machine or he connects a
>   machine.

In this section we describe how the program, the simulation
N.L.U.S., accepts these sentences and produces a textual
equivalent of an ACD for the problem under consideration. The
program is organised as shown in Figure 2.

INPUT/OUTPUT is the communication link between the user and
the computer. It accepts English input typed by a user in the
form of option-choice and true-false answers, or entity and
activity definitions, or normal sentences describing an action
which takes place in the problem under investigation. In the
first two cases, it checks for meaningless answers and in the
last case converts the sentence into a list of words to be
analysed afterwards. It also provides responses from the
system to the user in the form of option-choice menus, true-
false questions or descriptions of the problem in an ACD like
format.

The ANALYSER is the main coordinator of the language
understanding process. It handles the basic units of an
English sentence which describe an action i.e. the action
verb, the entity undertaking the action, and the other
entities involved (if any) in the action. It looks up words
in the dictionary, and performs morphemic analysis (e.g
realizing that "washes" is "es" added to the word "wash" and
modifying the dictionary accordingly).

The DICTIONARY consists of two parts. First, a closed
vocabulary which consists of standard English words (e.g the,
under, or, is etc) which are classified into categories
(determiners, prepositions etc). Second, an open vocabulary
of verbs and nouns listed independently. The list of verbs
consists of irregular verbs and a few verbs that can not be
analysed by the ANALYSER plus a few verbs which look different
but imply similar meanings (e.g the verbs 'leave', 'depart'
and 'exit'). Each time a new problem is started, the open
vocabulary is used as a basis to which the ANALYSER adds new
verbs (the root and the various forms used) and new nouns
(the singular form).

The CREATOR is a collection of routines which modify and build
a description of the problem based on the list of words with
their definitions returned by the ANALYSER. This is achieved
by storing a word (with a specific definition or definitions)
in the appropriate list in the COMPUTER REPRESENTATION of the
PROBLEM.

The COMPUTER REPRESENTATION of the PROBLEM is built by the

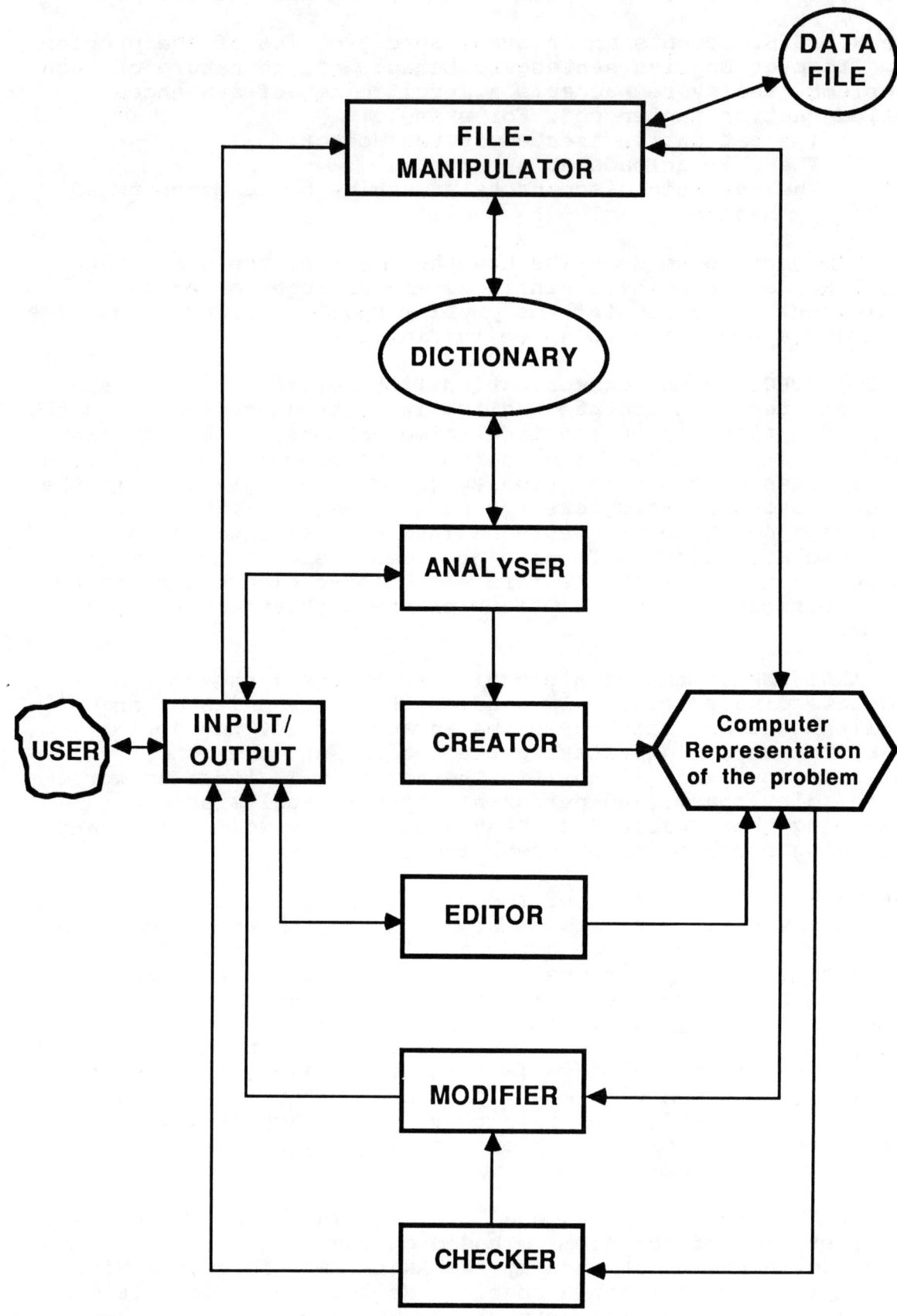

Fig. 2.   Organisation of the N.L.U.S.

system using the information it obtains from the CREATOR. This description consists mainly of logical lists of entities and activities and their relationships. The COMPUTER REPRESENTATION of the PROBLEM provides the data required for the system to generate the appropriate questions to be answered by the user in order to continue with the session. The data file containing the partially complete or complete session derives from this part of the system.

The CHECKER is the part of the system which scans through the partially built description of the problem for missing information. This occurs when either a life cycle is incomplete or when an entity has been mentioned but not been investigated. The CHECKER generates the appropriate question in each case.

The MODIFIER is a collection of routines which produce an English like description of the whole problem or specific parts of the problem. It consists of entity and activity descriptions.

The EDITOR is a debugging facility which allows the user to make changes to the problem, whether finished or not. Changes can be made to a specific life cycle or to the problem as a whole. Based on these changes the EDITOR modifies the model description and CHECKER moves the conversation forward accordingly.

The FILE-MANIPULATOR enables the user to save a problem in a data file and to recall it later to make changes or to continue the conversation from where it was stopped. To save a session, the program converts model representation lists and the open vocabulary to data file lists. To recall an old session, the program converts the data file lists into their original form.

5.   EXPERIMENTAL DESIGN AND STATISTICAL ANALYSIS OF THE MODELLING

Problem solving by stochastic process simulation is a statistical experiment. The analyst is attempting to obtain meaningful answers from a random process as efficiently as possible. Leaving aside the issue of model verification and validation, we shall look at some of the statistical issues involved.

There is a great danger when using a simulation model to make a single simulation run of somewhat arbitrary length and then treat the resulting simulation estimates as being the correct answers for the model. Since these estimates are random variables which may have large variances, they could differ greatly from the corresponding true answers. One reason for the historical lack of definitive data analysis is that simulation output data are rarely independent. Classical statistical analysis based on independent identically distributed observations are not directly applicable. A further difficulty in obtaining accurate estimates of a model's outputs is the computer cost of collecting the necessary amount of simulation output data.

There are two types of simulation with regard to analysis of
output. A terminating simulation is one in which the desired
measures of system performance are defined relative to when
some specific event occurs. For example, a battle is
completed when one side withdraws or is defeated. In this
case, the simulated time can often be a random variable. A
steady-state simulation is one for which the measures of
performance are defined as the length of the simulation goes
to infinity. Hence the length of the simulation must be large
enough to get good estimates of the quantities of interest.
The difficulty with steady-state simulation is recognising
when the simulation has arrived in the steady-state. Leading
output variables of interest can be calculated at regular
intervals of time to detect when the average of these draws to
a limit.

An obvious need in simulation output is a confidence interval
on the variables of interest. One method is to use a fixed
sample size approach. However, the simulator then has no
control after the sample size has been selected. Law and
Kelton [14] describe various methods for obtaining confidence
intervals with a specified precision. Among these are
sequential procedures by which the length of the simulation is
sequentially increased until an acceptable confidence interval
can be constructed using one of several techniques for
stopping the simulation run.

Similiar observations can be made when using a simulation
model to compare alternative systems. Many methods have been
proposed to achieve such comparisions, but the definitive
solution has not yet been given.

Variance reduction techniques are a method of conducting a
simulation experiment where the confidence in the model is
balanced against the cost of obtaining the solutions.
Efficient methods of obtaining desired levels of confidence
enable tighter levels of confidence to be sought at the same
cost. Methods for achieving these ends include common random
number streams, antithetic variates, control variates,
indirect estimation, and the use of conditional expectations
where appropriate.

Some simulation experiments deal with problems where the
structure of the goal of the study is less evident. In this
context, experimental design provides a way of deciding
beforehand which particular system variants to simulate so
that the desired information can be obtained at minimal cost.
Carefully thought out experiments are much more efficient than
a hit-or-miss sequence of runs which simply try a number of
systems unsystematically to see what happens. Factorial
designs and fractional factorial designs are particularly
useful in the early stages of experimentation when seeking the
factors which are important and how they might affect the
response. A whole variety of techniques known as response-
surface methodologies can be used as more is understood about
the behaviour of the model to find the optimal combination of

factor specifications.

Clearly the statistical aspects of simulation modelling can be
formidable. Coming at the end of the modelling process, there
is a great temptation to use the model wishfully under time
and resource constraint pressure. Whilst all of the
statistical aspects of the process have not been solved, there
is a large if not over-burdgeoning body of available
expertise. Hence in Figure 1 an intelligent output analyser
under the guidance of the analyst is seen as a desirable if
not necessary component of a simulation modelling environment.

6.   CONCLUSIONS AND FUTURE RESEARCH

What is clear to us from the development of AUTOSIM and LIBSIM
is what constitutes a sound simulation modelling environment.
The technical task of turning a formal model description of a
problem into a working experimental computer simulation
program should be a minor part of the overall problem solving
process. Problem definition and understanding, model
verification and model confidence, experimental design in
using the computer model and implementation of results are
currently difficult expert endeavours. The overall problem
solving process can be made much more efficient if the
technical part of the process is executed with minimum effort.
Our research, teaching and consulting activities suggest that
this is best achieved (relative to those methods available)
using an effective ISPG, which writes code in a well supported
widely available high level programming language, and in a
programming structure that is readily understood and
explicable to customers and other analysts alike. AUTOSIM and
LIBSIM constitute such a desirable working environment. They
contribute to an efficient life cycle for a project.
Furthermore, this environment encourages and easily permits
model updating through its relative simplicity and ease of
use. Too often a modelling environment encourages an analyst
to 'make do' with the current version of the program because
the effort involved in modifying it is too expensive and too
slow.

The complexity of problem that can be handled by the N.L.U.S.
has been steadily improved. Although only tested on
experimental problems, the N.L.U.S. can handle complex entity
characteristics and conditions for action. Output from the
N.L.U.S. has been used with AUTOSIM to successfully generate a
working computer model. A description of how this is achieved
will be given by Doukidis and Chew [9]. Hence the N.L.U.S. is
part of a suite of systems that enable a suitable problem to
be formulated as a model, the particular attributes of the
problem to be added, the running and reporting characteristics
of the program to be determined, and the computer simulation
program thence to be written.

The utility routines used in the N.L.U.S. have been
reassembled to provide the basis for a skeletal expert system
called ASPES which is described by Doukidis and Paul [12].
This system is currently used to teach undergraduate and
postgraduate students at the School as well as to aid

research. For example, a diagnostic expert system that
handles runtime and reporting errors has been developed for
the simulation system itself. A fully comprehensive error
message facility within a simulation program requires the
anticipation of all possible abuses of the system. Such
abuses usually occur in the subsequent editing of the
generated program. We have looked at the possibility of
adding more extensive diagnostics to LIBSIM, but this would be
at the expense of computer speed and memory. Therefore we
developed the expert system to help the user with runtime
errors and with simulation results that are obviously wrong.
This latter work is described by Doukidis and Paul [13].

Although the N.L.U.S. achieves the operating task set for it,
textual discourse is still slow and somewhat irritating.
Since computer based speech interpretation is not commercially
available, the next best computer aid to problem formulation
is probably graphical rather than using sentence reading and
sentence input. A simple system that enables an ACD to be
drawn using the output from the N.L.U.S. system has been
written and this system includes editing facilities. A visual
check on model formulation is therefore available, and
reformulated problems can still be submitted to the I.S.P.G.
as before. The graphical system is being enhanced to allow
the user to draw the ACD from scratch as well. This
experimental work, to be described by Paul [15], is the
forerunner to research into a system which is expected to use
a mixture of graphics, icons, menus and textual communication
as the front end to a fully automatic model formulator and
program generator.

The support for output analysis is limited at the moment. We
are working on a comfortable user-interface to widely
available statistical packages as part of an integrated
simulation support environment. Experience in constructing
AUTOSIM will be applied to the production of such an
intelligent interface. The statistical design and analysis
issues specific to stochastic process simulation, including
the proper employment of variance reduction techniques, the
determination of appropriate run-lengths and the avoidance of
bias due to transients will be accomodated within the systems.
Some work in this direction is in hand [20].

REFERENCES

1.   D.W.BALMER and R.J.PAUL (1986), CASM - The right
      environment for simulation. Journal of the Operational
      Research Society 37, 443-452

2.   A.BARR and E.A.FEIGENBAUM (1982), The Handbook of
      Artificial Intelligence, Vol. 2. Pitman, London.

3.   J.E.BEASLEY and G.WHITCHURCH (1984), O.R. education - a
      survey of young O.R. workers. Journal of the Operational
      Research Society 35, 281-288

4.   S.T.CHEW (1986) Program generators for discrete event

digital simulation modelling. Ph.D. Thesis, University of London.

5.   D.P.CHRISTY and H.J.WATSON (1983), The application of simulation: a survey of industry practice. Interfaces 13, 47-52

6.   A.T.CLEMENTSON (1982), Extended Control and Simulation Language. Cle. Com Ltd., Birmingham, U.K.

7.   J.G.CROOKES, D.W.BALMER, S.T.CHEW and R.J.PAUL (1986), A three phase simulation system written in Pascal. Journal of the Operational Research Society 37, 603-618

8.   A.A.R.EL SHEIKH, R.J.PAUL, A.S.HARDING and D.W.BALMER (1985), A microcomputer based simulation study of a port. CASM Report, Department of Statistics, L.S.E.

9.   G.I.DOUKIDIS and S.T.CHEW (1986), Automated simulation modelling, CASM Report, Department of Statistics, L.S.E. In preparation

10.   G.I.DOUKIDIS and R.J.PAUL (1985), Research into Expert Systems to aid Simulation Model Formulation. Journal of the Operational Research Society 36, 319-326.

11.   G.I.DOUKIDIS and R.J.PAUL (1986), Experiences in automating the formulation of discrete event simulation models. In A.I. Applied to Simulation (E.J.R.KERCKHOFFS, G.C.VANSTEENKISSE and B.P.ZEIGLER, Eds.) Simulation Series Vol.18, No.1. The Society for Computer Simulation, San Diego, U.S.A.

12.   G.I.DOUKIDIS and R.J.PAUL (1986), ASPES - a skeletal Pascal expert system, In Expert Systems and Artificial Intelligence in Decision Support Systems (C.A.Th.TAKKENBERG Ed.) Reidel, Dordrecht, Holland. In press

13.   G.I.DOUKIDIS and R.J.PAUL (1986) SIPDES : a simulation program debugger using an expert system. CASM Report, Dept. of Statistics, L.S.E.

14.   A.M.LAW and W.D.KELTON (1982), Simulation Modeling and Analysis. McGraw-Hill, New York.

15.   R.J.PAUL (1986), Graphical simulation model formulation, CASM Report, Department of Statistics, L.S.E. In preparation

16.   R.J.PAUL and S.T.CHEW (1986), Simulation modelling using an interactive simulation program generator. CASM Report, Dept. of Statistics. L.S.E.

17.   R.J.PAUL and G.I.DOUKIDIS (1986), Further developments in the use of artificial intelligence to formulate simulation problems. Journal of the Operational Research Society 37, 787-810

18.   M.PIDD (1984), Computer Simulation in Management Science.
      Wiley, Chichester

19.   T.G.POOLE and J.Z.SZYMANKIEWICZ (1977) Using Simulation
      to Solve Problems. McGraw-Hill, London.

20.   D.REYNIERS (1985), Automatic transient detection,
      confidence interval generation and run length control.
      Simulation project report, Dept. of Statistics, L.S.E.

8 Intelligent Front End to Box Jenkins Forecasting

David P. Reilly and Ana I. Timberlake

Automatic Forecasting Systems Inc., Drexel and Penn State Universities, and Timberlake Clarke Ltd

ABSTRACT

ARIMA and Transfer function models are extremely useful for analysing and forecasting time series data. This paper discusses a software system - AUTOBOX (version 1.02) - which incorporates the three major steps in the model building - identification, estimation and forecasting - as a complete automatic feature, performed without any user intervention. AUTOBOX also offers the intervention detection techniques.

The authors:

David P. Reilly is the founder of Automatic Forecasting Systems, Inc. and a lecturer at Drexel and Penn State Universities in the U.S.. David has over 20 years experience in the use of time series techniques, specifically the Box-Jenkins methods and has developed the AUTOBOX algorithms. AUTOBOX has been available on the Computer Sciences and Chase Econometrics bureaux for 10 years.

Ana I. Timberlake is the Managing Director and a Senior Consultant of Timberlake Clark Ltd, a management consultancy specialising in statistical and economic applications and software. Ana has over 20 years of consulting experience in the area of statistical analysis.

INTRODUCTION

Analysis of time series data is, and has been, an important
concept in a wide variety of fields. Over the years a number of
quantitative techniques have been developed for analysing and
predicting time series values. These methods generally involve
forming an equation that models the history of the time series,
the rationale being that if past values follow a representative
pattern, then it stands to reason that future values might do the
same. Therefore, the first task of the time series analyst is
concerned with the identification of an equation which represents
the history of the time series. This identification is time
consuming and sometimes problematic. The time series analyst
must often rely on judgemental factors and assume the model form
and the more sophisticated the model is, the more it requires
certain level of technical expertise and the commitament of the
analysts time. A computerised procedure, such as the algorithms
discussed in this paper, for determining the appropriate model
reduces these barriers and widens the range of analytical tools
available to the time series analyst.

The automatic modelling algorithms under discussion utilizes G.
Box and G. Jenkins (1976) models. The Box and Jenkins method is
a comprehensive statistical procedure for deriving a model from
the information in the data itself.

"BOX-JENKINS"

Box and Jenkins (1976) discuss two model types which describe a
single time series.

1- <u>ARIMA (Autoregressive Integrated Moving Average)</u>. This type
of model is termed univariate as the single series is modelled as
a function of its own past values. The optional ARIMA model for
any time series can be derived from the general equation:

$$O_p(B) \ (Y_t - u) = O_o + O_q \ (B)A_t$$

where

Y_t = the discrete time series
u = the mean of the stationary series
$-$ = the differencing factor(s)
O_p = the autoregressive factor(s)
O_o = the deterministic trend
O_q = the moving average factor(s)
A_t = the noise series

and

B = the backshift operator

The second type of model describes a single time series as a
function of its own past values and the values of one or more
independent input series. These models are called TRANSFER

FUNCTION models, as the information in the input series is transfered to the dependent output series. These models are more useful than ARIMA models when the single series is a function of some related time series. Again, the optional transfer function model can be derived from the general equation:

$$Y_t = f1(X1_t) + f2(X2_t) + \ldots fN(XN_t) + fA(A_t)$$

where

Y_t = the dependent output series

$X1_t$ = independent input series 1, (independent of output)

$X2_t$ = independent input series 2

XN_t = independent input series N

A_t = the noise series

$f1$ = the transfer function between series Y and series X1

$f2$ = the transfer function between series Y and series X2

fN = the transfer function between series Y and series XN

fA = the noise model

Where A_t is the residuals from the transfer function between Y and X, labelled as the noise series and modelled by an ARIMA equation. The combined transfer function-noise model is often referred to simply as a transfer function.

There are three distinct model building stages for both the ARIMA and TRANSFER FUNCTION model types,

1. IDENTIFICATION (or specification) of forecasting models involving the use of rough data analysis tools (range-mean plots, auto-correlation (ACF) and partial auto-correlation functions (PACF) to arrive at initial guesses of the data transformations, degrees of differencing needed to induce stationarity and the degrees of the polynomials appearing in the various auto-regressive and moving average operators appearing in the model.

2. Estimation of Diagnostic Checks (or fitting). Estimation involves using fully efficient (likelihood) methods for estimating the parameters of the model, their standard errors and correlations, and the residual variances and co-variances. Since no model can ever be 'correct', checking is an important part of this step which involves looking for model inadequacies or for areas where simplification can take place. The most important model criticism criteria are:

- the residuals left unexplained by the model - are there any abnormally large residuals which can be linked to known external factors or other explanatory variables?

- the residual <u>correlations</u> and <u>partial correlations</u>- do they provide evidence that the model can be elaborated in any particular direction?

3. <u>Forecasting</u>. Use the model to generate forecasts and confidence limits for future values of the time series.
<u>THE "AUTOBOX" SOFTWARE PACKAGE</u>

The software package AUTOBOX incorporates both Box-Jenkins model types
* ARIMA with or without Intervention Detection
* TRANSFER FUNCTION by cross-correlation or common filter methods

in a completely automatic mode. It should be noted though that one can still choose to use the programme in an non-automatic way as with other conventional packages. The AUTOBOX system follows a tree structure as described in Fig. 1.

The AUTOBOX programme, developed in FORTRAN, is menu-driven with a very convenient question and answer format. HELP options are available at various stages of the process and high-resolution graphics are obtainable for micros using the right configurations. AUTOBOX runs on PC-compatible computers with PC-DOS or MS-DOS and requires 320K of RAM. Although the 8087 math-coprocessor is not mandatory, the programme makes use of it when available with a noticeable improvement in speed (processing is on average 10 times faster when the math-coprocessor is available).

AUTOBOX is a product of AFS Inc (Automatic Forecasting Systems Inc), a firm of statistical consultants based in Pennsylvania. They have also developed other forecasting systems, such as MTS, for multivariate Box-Jenkins. In the UK, AUTOBOX is supported and marketed by Timberlake Clark Ltd.

<u>COMPUTATIONAL METHODS AND ALGORITHMS IN AUTOBOX</u>

Both the ARIMA and TRANSFER FUNCTION model types involve a three stage iterative process

* Identification
* Estimation and diagnostic checking
* Forecasting

1. <u>ARIMA</u>   Fig. 2 portrays the steps in the ARIMA algorithm. The first step is the "identification" of the model. The aim of this process is to make the time series stationary - both the mean and variance should be constant over time. As there are two types of non-stationarity, there are two methods to induce it. Applying the appropriate differencing factors to a series - given by the autocorrelation patterns - creates a "mean" stationary series; applying the correct power transformation (lambda value) - by

Figure 1. THE AUTOBOX SYSTEM

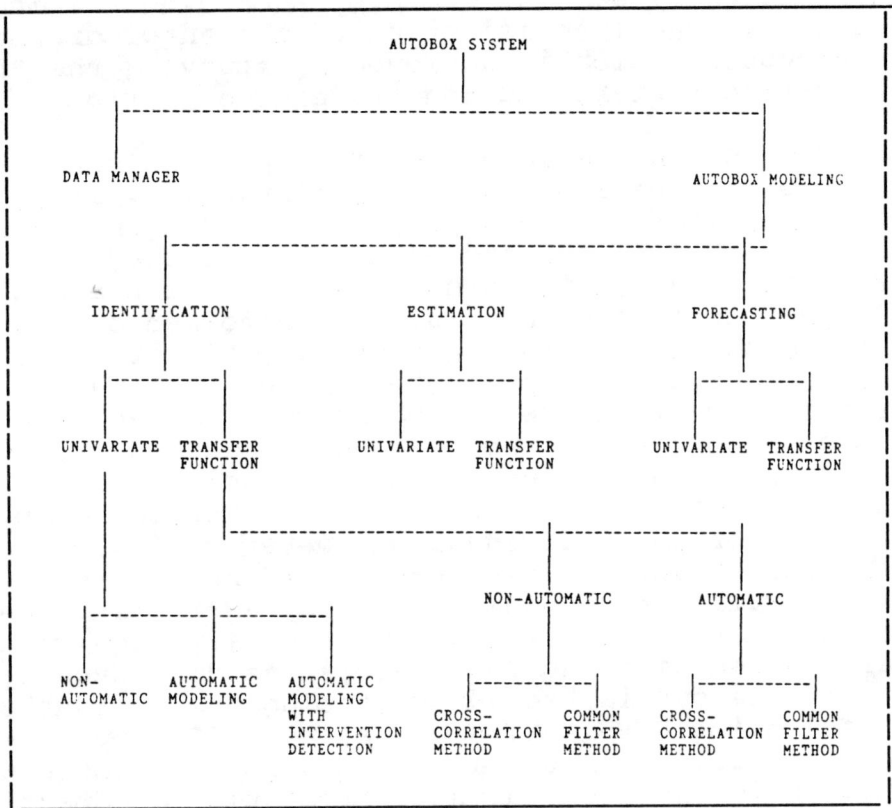

Figure 2. ALGORITHM FOR THE ARIMA MODEL

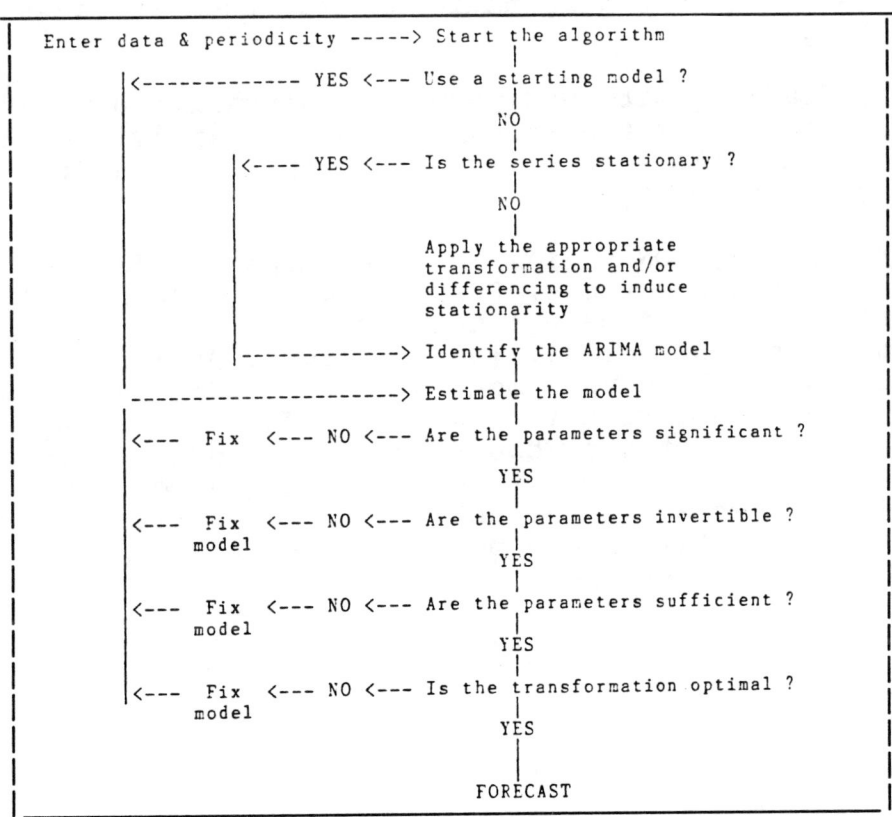

using the Box-Cox [1] error sum of squares test- creates a variance stationary series. The step succeding the inducement of stationarity is the identification of the autoregressive/moving average structure which is achieved by studying the "patterns" of the appropriate differenced and transformed series.

The second phase in the model building process consists of the estimation of the parameters in the tentatively identified model. The non-linear least squares procedure, based on the Marquard algorithm [2- pp 504-505] and backcasting procedures ([2]- pp 212-220) are used to estimate these parameters. There are then three basic diagnostic checks that must be performed on the estimated model. These checks are the "necessity", "invertibility" and "sufficiency", i.e. each parameter included in the model should be statistically significant (necessary) and each factor must be invertible. In addition, the residuals obtained from the estimation of the model should be white noise (model sufficiency). The test for "necessity" is performed by examining the t-ratios for the individual parameter estimates - parameters with non-significant coefficients should be deleted from the model; "invertibility" is determined by extracting the roots from each factor in the model - the invertibility requirements of the model are tested with the Shur theorem as discussed by Cryer [3]; model "sufficiency" is tested by testing the residuals for white noise - following Box-Jenkins guidelines ([2]- pp 392-395) - in much the same way as in the identification process and, if there are patterns in the residual autocorrelation and partial autocorrelation, new parameters are added to the model. Finally, the optimality of the power transformation is tested [1]. If the model passes all the diagnostic checks, then it is suitable for forecasting.

Model forecasting with the properly identified and estimated model is simply a mathematical process of applying the model form to the actual time series data and computing the forecast values from a given time origin.

2. TRANSFER FUNCTION - Fig. 3 and 4 shows the flow charts for the transfer function options. Before starting the modelling process, the analyst must define which independent input variables are driving the output series. The transfer function identification phase involves the development of a univariate ARIMA model for each of the time series in the model, following the ARIMA algorithm. The next step consists of the computation of certain key statistics which are ,in turn, examined for clues as to the model form, much as in the univariate identification. Having tentatively identified the model form, one can proceed to the next phase - estimation.

Figure 3. ALGORITHM FOR THE TRANSFER FUNCTION MODEL
(USING THE PREWHITENING/CROSS-CORRELATION METHOD)

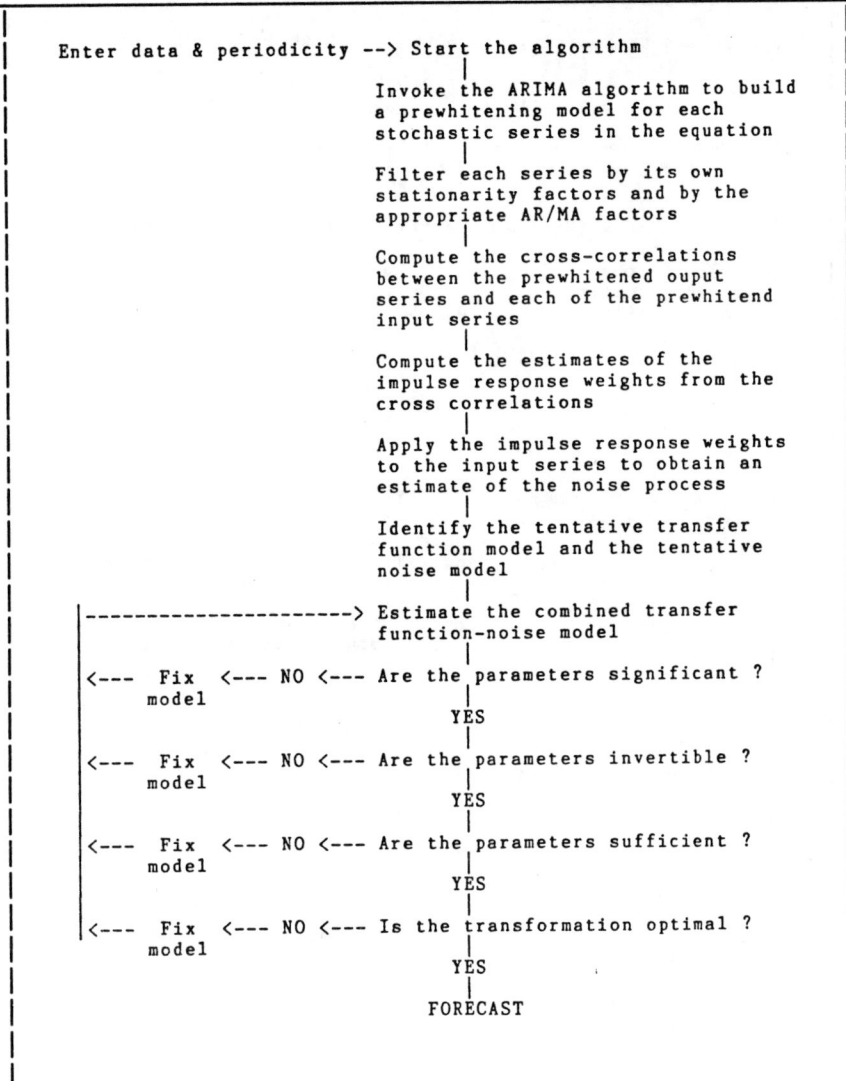

In the identification phase, four distinct methods may be chosen,

* Prewhitening/Cross-Correlation Method
* Common Filter/Least Squares Method (2)

This paper will only cover, in some detail the first model type - Prewhitening/Cross-Correlation Method. Figure 3 shows the flow chart for this algorithm. Figure 4 shows the flow chart for the Common Filter/Least Squares Method.

The procedure for Transfer Function model identification outlined by Box-Jenkins uses the cross-correlation between two prewhitened series to tentatively identify the model form. The first step to this process is to use the ARIMA model development algorithm described above, for each time series in 'the equation', i.e. each series must be made stationary by applying the appropriate differencing and transformation parameters from its ARIMA model.

**Figure 4 . ALGORITHM FOR THE TRANSFER FUNCTION METHOD
(USING THE COMMON FILTER/LEAST SQUARES METHOD)**

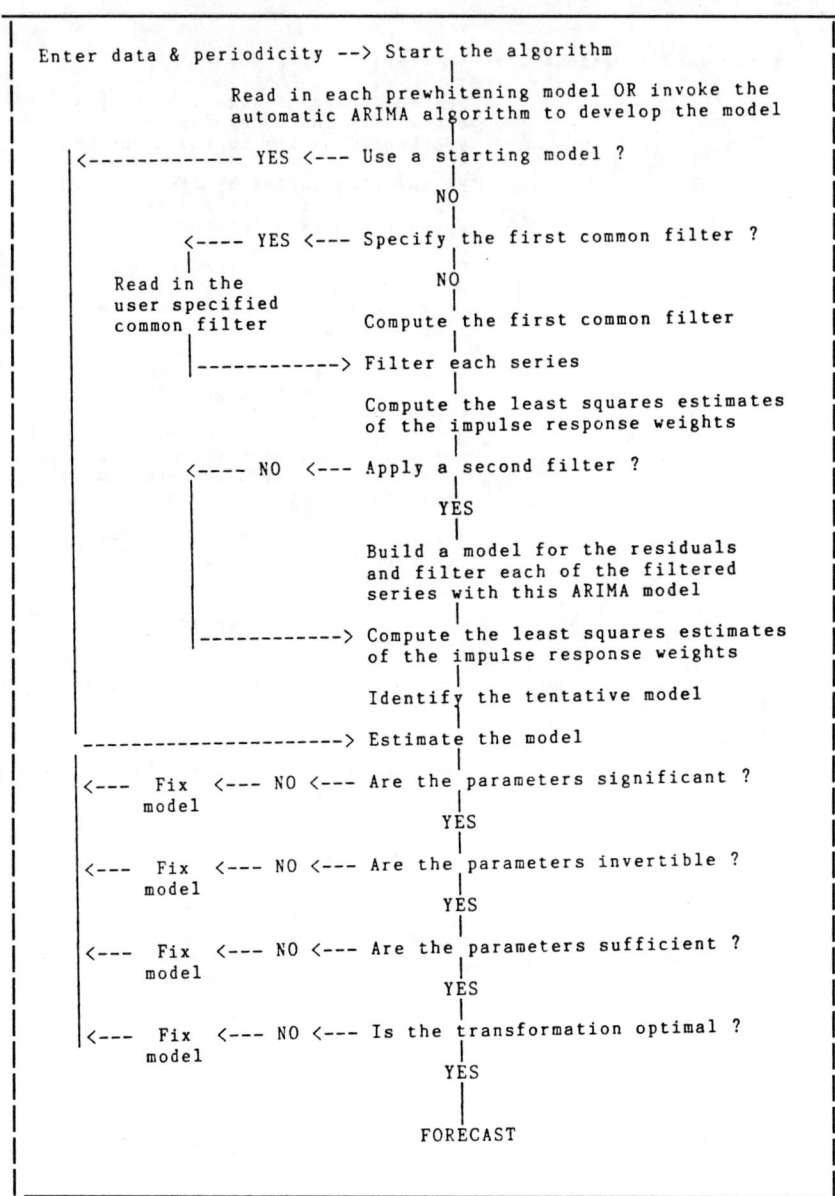

The stationary time series are, in turn, "prewhitened".
Prewhitening is important on two accounts. First, it is
necessary to induce stationarity for each stochastic series in
the equation. Second, filtering the input and output series
removes any interrelationship so that the cross-correlation
function reveals only the interrelationships.

The tentative transfer function model is identified by examining
the pattern in the estimated impulse response weights, as per the
rules stated by Box and Jenkins ([2], pp 346-353). The tentative
noise model and its preliminary parameter values are obtained by
inputing the estimated noise series to the ARIMA algorithm.
Starting values for the coefficients in the transfer model are
computed using the routine listed in Box-Jenkins ([2], pp 511-513).

These preliminary parameter estimates are then input to a routine that computes the simultaneous parameter estimates via the Marquardt non-linear least squares estimation algorithm. The standard errors of the parameter estimates are examined for necessity with the student t test. The invertibility requirements of the model are tested with the Schur theorem as discussed by Cryer [3]. Model sufficiency checks include tests for both the transfer model and the noise model. The residuals are examined for structure following the Box-Jenkins guidelines ([2], pp 392-395). Finally, the optimality of the power transformation is tested [1]. If the model passes all of the diagnostic checks, then it is suitable for forecasting.

CONCLUSION

The programme AUTOBOX represents a significant advance over programmes that are not automatic. For the casual Box-Jenkin user, the automatic procedure provides a systematic approach for those common cases in which an initial inspection of the ACF and PACF fails to indicate a clear choice among possible successor models and/or alliviates greatly the analysts time in defining a feasible model. One can also envisage a number of potential applications, such as military systems, for which a fast on-line Box-Jenkins identification procedure should be extremely useful.

REFERENCES

1.   Box, G.E.P., and Cox, D. R. (1964), "An analysis of Transformations" <u>Journal of the Royal Statistical Society</u>, B26, 211-252.

2.   Box, G.E.P., and Jenkins, G.M. (1976), <u>Time Series Analysis: Forecasting and Control</u> (Revised Edition), Holden Day.

3.   Cryer, J.D. (1986), <u>Time Series Analysis</u>, Duxbury Press.

4.   McCleary, R. and Hay, R. (1980), <u>Applied Time Series Analysis for the Social Sciences</u>, Sage.

5.   Reilly, D.P. (1986), <u>AUTOBOX User's Guide</u>, Automatic Forecasting Systems Inc, P.O. Box 563, Pennsylvania, USA.

6.   Reilly, D.P. (1980), "Recent Experiences with an Automatic Box-Jenkins Modelling Algorithm", <u>Time Series Analysis</u>, eds. O.D. Anderson and M.R. Perrymanan, North-Holland, pp 493-508.

7.   Shumway, R.H. (1986), "AUTOBOX (Version 1.02), Statistical Computing Software Review, <u>The American Statistician</u>, Vol. 40, No. 4, pp 299-300.

APPENDIX 1 : EXAMPLE OF THE AUTOMATIC ALGORITHM IN PROCESS

We now present an example of how the automatic algorithm modelled
the attic temperature data that was originally analysed by
Downing and Pack (1982). The AUTOBOX program (Reilly, 1984),
which houses the automatic transfer function algorithm, produces
an output file that optionally lists the various steps taken
during model development. The following material contains
excerpts from this listing and a brief discussion of the entire
modelling process.

The first step to build an ARIMA model for each of the time
series in the equation. The AUTOBOX program allows the user to
either specify the model form or to have the program develop it
automatically. We chose to have both the input series and the
output series modelled automatically. These are the ARIMA models
that the program developed.

The ARIMA model for the input series:

$$(1-1.25B+.34B_2+.45B^3) \; (Y_{t}-24.3) = A^t$$

As an aside, we note here that Downing and Pack developed an
ARIMA model for the input and the output series. Since their
ARIMA model contained a differencing factor, this differencing
factor was used to prewhiten the output series. The AUTOBOX
program also prewhitens the output series by the
autoregressive/moving average factors of the input series; but,
it does not assume that the differencing that is necessary for
the input is also needed for the output series, for the
differencing factors from this model are used to pre-whiten the
output series. We found that if Downing & Pack had not
differenciated the output series during prewhitening, then their
"vanishing" transfer function would have appeared.

EXAMPLE OF THE AUTOMATIC ALGORITHM IN PROCESS

We now present an example of how the automatic algorithm modeled
the attic temperature data that was originally analyzed by
Downing and Pack (1982). The AUTOBOX program (Reilly, 1984),
which houses the automatic transfer function algorithm, produces
an output file that optionally lists the various steps taken
during model development. The following material contains
excerpts from this listing and a brief discussion of the entire
modeling process.

The first step to build an ARIMA model for each of the time
series in the equation. The AUTOBOX program allows the user to
either specify the model form or to have the program develop it
automatically. We chose to have both the input series and the
output series modeled automatically. These are the ARIMA models
that the program developed:

The ARIMA model for the input series:

$$(1-1.25B+.34B^2) \ (1-.39B^{24}) \ (X_t-18.6) = (1+.4B^{48})A_t$$

The ARIMA model for the output series:

$$(1-1.13B-.23B^2+.45B^3) \ (Y_t-24.3) = A_t$$

As an aside, we note here that Downing and Pack developed an
ARIMA model for the input series which they then used to filter
both the input and the output series. Since their ARIMA model
contained a differencing factor, this differencing factor was
used to prewhiten the output series. The AUTOBOX program also
prewhitens the output series by the autoregressive/moving average
factors of the input series; but, it does not assume that the
differencing that is necessary for the input is also needed for
the output. That is why it builds an ARIMA model for the output
series, for the differencing factors from this model are used to
prewhiten the output series. We found that if Downing and Pack
had not differenced the output series during pre-whitening, then
their "vanishing" transfer function would have appeared.

Continuing with the automatic modeling process, the next step is to identify a tentative transfer function model. This entails prewhitening the input and output series, and then computing the estimates of the impulse response weights. The impulse response weights are then examined for clues as to the appropriate transfer function model form. The AUTOBOX program shows this phase of the process in the following report:

```
-----------------------------------------------------
-----------------------------------------------------
STEP   1 OF TRANSFER FUNCTION IDENTIFICATION -- THE ANALYSIS BETWEEN

INPUT SERIES  1 : OUT.DAT
OUTPUT SERIES   : IN.DAT
-----------------------------------------------------
-----------------------------------------------------


                    THE PREWHITENING MODEL
                    ======================
*********************************************************
DATA : OUT.DAT                              168 OBSERVATIONS    <-- The filter
                                                                    applied
DIFFERENCING FACTORS : NONE                                         to the
                                                                    input
BACKCASTING  : OFF                                                  series.
*********************************************************
PREWHITENING MODEL PARAMETERS
*********************************************************
                TYPE   LAG COEFFICIENT

*********************************************************
 1 MEAN                     .18589E+02
 2 AUTOREGRESSIVE 1     1    .12524E+01
 3 AUTOREGRESSIVE 1     2   -.34113E+00
 4 AUTOREGRESSIVE 2    24    .39003E+00
 5 MOVING AVERAGE 1    48   -.40099E+00
*********************************************************


                    THE PREWHITENING MODEL
                    ======================
*********************************************************
DATA : IN.DAT                               168 OBSERVATIONS    <-- The filter
                                                                    applied
DIFFERENCING FACTORS : NONE                                         to the
                                                                    output
BACKCASTING  : OFF                                                  series.
*********************************************************
PREWHITENING MODEL PARAMETERS
*********************************************************
                TYPE   LAG COEFFICIENT

*********************************************************
 1 MEAN                     .24347E+02
 2 AUTOREGRESSIVE 1     1    .12524E+01
 3 AUTOREGRESSIVE 1     2   -.34113E+00
 4 AUTOREGRESSIVE 2    24    .39003E+00
 5 MOVING AVERAGE 1    48   -.40099E+00
*********************************************************



                 CROSS-CORRELATION ANALYSIS
                 ==========================

        MEAN OF THE INPUT SERIES        :    .91275E-03
        STANDARD DEVIATION              :    .14311E+01

        MEAN OF THE OUTPUT SERIES       :   -.38621E-03
        STANDARD DEVIATION              :    .41472E+00

        NUMBER OF OBSERVATIONS          :     142

    .
    .
    .
```

```
------------------------------------------------------------------------
                      THE IMPULSE RESPONSE WEIGHTS
                      ==============================

      IMPULSE RESPONSE WEIGHTS (REGRESSION WEIGHTS) ARE PROPORTIONAL TO
      THE CROSS-CORRELATIONS VIA THE RATIO OF STANDARD DEVIATIONS (SD)

            V(0)  =   CC(0)  *  ( SD OF OUTPUT) / ( SD OF INPUT )
        .49751E-01 =     .172 * (   .41472E+00 ) / (  .14311E+01 )


              :-------------------------------------------------:
              : LAG  : IMPULSE RESPONSE WEIGHT : T-RATIO :
              :-------------------------------------------------:
              :V(  0) :        .49751E-01        :    2.0313:
              :V(  1) :        .35678E-01        :    1.4515:
              :V(  2) :        .45117E-01        :    1.8289:
              :V(  3) :        .39508E-01        :    1.5957:
              :V(  4) :        .48573E-01        :    1.9547:
              :V(  5) :        .84003E-01        :    3.3680:
              :V(  6) :        .75535E-02        :     .3017:
              :V(  7) :        .33739E-01        :    1.3427:
              :V(  8) :        .39265E-01        :    1.5567:
              :V(  9) :        .12414E-01        :     .4903:
              :V( 10) :       -.32753E-02        :    -.1289:
              :V( 11) :       -.12934E-01        :    -.5069:
              :V( 12) :        .32622E-01        :    1.2736:
              :V( 13) :       -.47305E-01        :   -1.8396:
              :V( 14) :        .65512E-02        :     .2538:
              :V( 15) :       -.28735E-01        :   -1.1086:
              :V( 16) :       -.57966E-01        :   -2.2274:
              :V( 17) :        .41742E-02        :     .1597:
              :V( 18) :       -.21747E-01        :    -.8289:
              :V( 19) :       -.37883E-01        :   -1.4380:
              :V( 20) :       -.46994E-02        :    -.1776:
              :V( 21) :       -.11860E-01        :    -.4464:
              :V( 22) :        .41447E-02        :     .1554:
              :V( 23) :        .19747E-01        :     .7370:
              :V( 24) :       -.12719E-01        :    -.4727:
              :V( 25) :        .16381E-01        :     .6062:
              :V( 26) :        .15564E-01        :     .5734:
              :V( 27) :        .29423E-01        :    1.0793:
              :V( 28) :        .33486E-01        :    1.2229:
              :V( 29) :        .32631E-02        :     .1186:
              :V( 30) :        .38413E-01        :    1.3902:
              :V( 31) :        .16342E-01        :     .5888:
              :V( 32) :        .23185E-01        :     .8314:
              :V( 33) :       -.14276E-01        :    -.5096:
              :V( 34) :       -.19845E-01        :    -.7050:
              :V( 35) :        .10094E-01        :     .3569:
              :V( 36) :       -.39435E-01        :   -1.3877:
              :V( 37) :       -.30897E-03        :    -.0108:
              :V( 38) :       -.17088E-01        :    -.5955:
              :V( 39) :       -.20729E-01        :    -.7189:
              :V( 40) :       -.40410E-02        :    -.1394:
              :V( 41) :       -.42779E-01        :   -1.4688:
              :V( 42) :       -.10614E-01        :    -.3626:
              :V( 43) :        .14684E-01        :     .4990:
              :V( 44) :       -.39561E-01        :   -1.3376:
              :V( 45) :       -.60045E-02        :    -.2020:
              :V( 46) :        .19850E-01        :     .6641:
              :V( 47) :       -.20075E-01        :    -.6680:
              :V( 48) :        .58732E-02        :     .1944:
              :-------------------------------------------------:


                 AUTOBOX SUGGESTED TENTATIVE MODEL FORM
                 =====================================

                 NUMBER OF OUTPUT LAG FACTORS =    1
                 NUMBER OF INPUT LAG FACTORS  =    1
                 DELAY (LAG PARAMETER)        =    0

      OUTPUT LAG FACTOR #1'S BACKORDER POWERS:    1
      AND ITS ASSOCIATED STARTING VALUES:     .87568

      INPUT LAG FACTOR #1'S BACKORDER POWERS:    0    1    2
      AND ITS ASSOCIATED STARTING VALUES:     .04975  .00789  -.01388
------------------------------------------------------------------------
```

Once the tentative transfer model is determined, the program computes the estimated noise process by applying the above impulse response weights to the original time series. This noise series is then modeled by an automatic ARIMA algorithm to obtain a preliminary noise model. The combined transfer-function – noise model is then estimated.

```
-------------------------------------------------------
           MEAN OF THE RESIDUAL SERIES      :   .14192E-01
           STANDARD DEVIATION               :   .29122E+01
           NUMBER OF OBSERVATIONS           :    120
           MEAN DIVIDED BY THE STANDARD
           ERROR OF THE MEAN                :   .53345E-01


                    THE AUTOCORRELATIONS
                    --------------------

LAGS   1-  8      .963    .865    .716    .523    .303    .071   -.158   -.368
STANDARD ERROR  (.091)  (.154)  (.190)  (.212)  (.222)  (.226)  (.226)  (.227)

LAGS   9- 16     -.546   -.682   -.769   -.802   -.782   -.709   -.592   -.440
STANDARD ERROR  (.232)  (.242)  (.258)  (.276)  (.295)  (.312)  (.325)  (.334)

LAGS  17- 24     -.264   -.075    .112    .285    .434    .548    .621    .649
STANDARD ERROR  (.338)  (.340)  (.340)  (.341)  (.343)  (.347)  (.354)  (.363)

LAGS  25- 32      .629    .566    .465    .333    .181    .019   -.143   -.290
STANDARD ERROR  (.373)  (.382)  (.388)  (.393)  (.395)  (.396)  (.396)  (.397)

LAGS  33- 40     -.416   -.515   -.580   -.606   -.594   -.543   -.460   -.350
STANDARD ERROR  (.398)  (.402)  (.407)  (.414)  (.422)  (.428)  (.434)  (.438)

LAGS  41- 48     -.219   -.075    .071    .209    .330    .427    .494    .528
STANDARD ERROR  (.440)  (.441)  (.441)  (.442)  (.442)  (.444)  (.448)  (.452)


                  THE PARTIAL AUTOCORRELATIONS
                  ----------------------------

LAGS   1-  8      .963   -.831   -.394   -.365    .039   -.031   -.037    .099
STANDARD ERROR  (.091)  (.091)  (.091)  (.091)  (.091)  (.091)  (.091)  (.091)

LAGS   9- 16     -.041    .046   -.147    .074   -.010    .160   -.128    .006
STANDARD ERROR  (.091)  (.091)  (.091)  (.091)  (.091)  (.091)  (.091)  (.091)

LAGS  17- 24      .051    .020   -.102    .031    .039   -.080   -.021   -.095
STANDARD ERROR  (.091)  (.091)  (.091)  (.091)  (.091)  (.091)  (.091)  (.091)

LAGS  25- 32     -.018    .042   -.032   -.003   -.040   -.056   -.004    .064
STANDARD ERROR  (.091)  (.091)  (.091)  (.091)  (.091)  (.091)  (.091)  (.091)

LAGS  33- 40     -.099   -.190   -.041    .053    .096   -.036   -.086   -.004
STANDARD ERROR  (.091)  (.091)  (.091)  (.091)  (.091)  (.091)  (.091)  (.091)

LAGS  41- 48      .083    .023   -.093   -.027    .050   -.134    .045    .021
STANDARD ERROR  (.091)  (.091)  (.091)  (.091)  (.091)  (.091)  (.091)  (.091)

-------------------------------------------------------
```

<-- The ACF
and the
PACF of
the est-
imated
noise
series.

AUTOBOX then estimates the revised model and reports the results of this estimation in the model form table shown here:

```
-----------------------------------------------------------------


            ESTIMATION OF THE TENTATIVELY IDENTIFIED MODEL FORM
            ===================================================
     ***********************************************************************
     DATA :   Y = IN.DAT                          168 OBSERVATIONS

     DIFFERENCING FACTORS : NONE

     BACKCASTING : OFF
     ***********************************************************************
     ***********************************************************************
     NOISE SERIES

     DIFFERENCING FACTORS ON NOISE : NONE
     ***********************************************************************
     NOISE MODEL PARAMETERS
     ***********************************************************************
                  FACTOR   LAG COEFFICIENT       T RATIO

     ***********************************************************************
      1 MEAN                   .24387E+02         129.96
      2 AUTOREGRESSIVE 1    1   .85955E+00          21.24
     ***********************************************************************
     ***********************************************************************
     INPUT SERIES   1
     ----------------

     DATA -   X1 = OUT.DAT

     DIFFERENCING FACTORS : NONE (ASSUMED MEAN OF SERIES =   .17767E+02 )

     VALUE OF LAG PARAMETER IS  0
     ***********************************************************************
     TRANSFER FUNCTION PARAMETERS
                  FACTOR   LAG COEFFICIENT       T RATIO

     ***********************************************************************
      3 OUTPUT LAG     1    1   .91172E+00          54.99
      4 INPUT LAG      1    0   .51863E-01          13.23
     ***********************************************************************


-----------------------------------------------------------------
```

After displaying the revised model, the automatic algorithm examines the residual autocorrelations, partial auto-correlations and cross-correlations. It finds that there is still some structure in the noise, and so it adds a parameter to the model. After a few iterations of adding parameters and checking for necessity, invertibility and sufficiency, AUTOBOX lists the final estimation of the identified model. Following this model form, the program lists the residual ACF, PACF and CCF as proof that the model is indeed adequate. The next two pages show this report.

The first pass estimation of the combined transfer function
- noise model, as shown in the AUTOBOX output report:

```
-------------------------------------------------------

            ESTIMATION OF THE TENTATIVELY IDENTIFIED MODEL FORM
            ===================================================
**********************************************************************
DATA :   Y = IN.DAT                           168 OBSERVATIONS

DIFFERENCING FACTORS (ORDER,DEGREE) : ( 1, 1)

BACKCASTING : OFF
**********************************************************************
**********************************************************************
NOISE SERIES

DIFFERENCING FACTORS ON NOISE : NONE
**********************************************************************
NOISE MODEL PARAMETERS
**********************************************************************
              FACTOR   LAG COEFFICIENT     T RATIO

**********************************************************************
  1 AUTOREGRESSIVE 1      1   .91343E+00       16.38
  2 AUTOREGRESSIVE 1      4  -.42184E-01        -.76
  3 MOVING AVERAGE 1      1   .10205E+01       34.83
**********************************************************************
**********************************************************************
INPUT SERIES  1
---------------

DATA -  X1 = OUT.DAT

DIFFERENCING FACTORS (ORDER,DEGREE) : ( 1, 1)

VALUE OF LAG PARAMETER IS  0
**********************************************************************
TRANSFER FUNCTION PARAMETERS
              FACTOR   LAG COEFFICIENT     T RATIO

**********************************************************************
  4 OUTPUT LAG    1      1   .92326E+00       38.77
  5 INPUT LAG     1      0   .38663E-01        2.38
  6 INPUT LAG     1      1   .11958E-02         .04
  7 INPUT LAG     1      2  -.11126E-01        -.59
**********************************************************************

-------------------------------------------------------
```

Diagnostic checking follows estimation, and so AUTOBOX
deletes all of the non-significant parameters. This is a
parameter by parameter process, since it is not certain
that the deletion of a parameter won't change the estimates of
the remaining parameters. After several passes through
estimation and diagnostic checking, AUTOBOX reports the
following model:

```
              ESTIMATION OF THE TENTATIVELY IDENTIFIED MODEL FORM
              ====================================================
**************************************************************************
DATA :    Y = IN.DAT                                    168 OBSERVATIONS

DIFFERENCING FACTORS (ORDER,DEGREE) : (  1,  1)

BACKCASTING : OFF
**************************************************************************
**************************************************************************
NOISE SERIES

DIFFERENCING FACTORS ON NOISE : NONE
**************************************************************************
NOISE MODEL PARAMETERS
**************************************************************************
                    FACTOR    LAG  COEFFICIENT      T RATIO

**************************************************************************
  1 AUTOREGRESSIVE 1       1    .93152E+00        87.55
  2 MOVING AVERAGE 1       1    .10311E+01       190.20
**************************************************************************
**************************************************************************
INPUT SERIES  1
---------------

DATA -   X1 = OUT.DAT

DIFFERENCING FACTORS (ORDER,DEGREE) : (  1,  1)

VALUE OF LAG PARAMETER IS  0
**************************************************************************
TRANSFER FUNCTION PARAMETERS
                    FACTOR    LAG  COEFFICIENT      T RATIO

**************************************************************************
  3 OUTPUT LAG       1       1    .90354E+00        41.36
  4 INPUT LAG        1       0    .49542E-01        11.01
**************************************************************************
```

Diagnostic checks on this model determine that all of
the parameters are significant (necessary), but, not all
of the parameters are invertible. Hence, AUTOBOX decides to
fix the model as reported below:

```
                        THE DIAGNOSTIC CHECKS
                        =====================


INVERTIBILITY CHECK : THE NOISE MODEL MOVING AVERAGE FACTOR IS NEARLY NON
                      INVERTIBLE.  THIS FACTOR WILL BE CANCELLED WITH
                      DIFFERENCING FACTORS ACROSS THE MODEL.
```

```
------------------------------------------------------------
                    FINAL ESTIMATION OF THE MODEL
                    -----------------------------
              (CONCLUDES THIS PHASE OF MODEL IDENTIFICATION)
    ************************************************************
    DATA :   Y = IN.DAT                            168 OBSERVATIONS

    DIFFERENCING FACTORS : NONE

    BACKCASTING : OFF
    ************************************************************
    ************************************************************
    NOISE SERIES

    DIFFERENCING FACTORS ON NOISE : NONE
    ************************************************************
    NOISE MODEL PARAMETERS
    ************************************************************
                   FACTOR   LAG COEFFICIENT      T RATIO

    ************************************************************
      1 MEAN                      .24385E+02      142.39
      2 AUTOREGRESSIVE 1    1     .81931E+00       16.82
      3 MOVING AVERAGE 1    2    -.17970E+00       -2.17
    ************************************************************
    ************************************************************
    INPUT SERIES  1
    ---------------


    DATA -   X1 = OUT.DAT

    DIFFERENCING FACTORS : NONE (ASSUMED MEAN OF SERIES =  .17767E+02 )

    VALUE OF LAG PARAMETER IS  0
    ************************************************************
    TRANSFER FUNCTION PARAMETERS
                   FACTOR   LAG COEFFICIENT      T RATIO

    ************************************************************
      4 OUTPUT LAG     1     1    .91185E+00       50.82
      5 INPUT LAG      1     0    .52030E-01       12.31
    ************************************************************
```

THE RESIDUAL STATISTICS
========================

```
    SUM OF SQUARES :   .17161E+02      DEGREES OF FREEDOM  :     161
    MEAN SQUARE    :   .10659E+00      NUMBER OF RESIDUALS :     166
    R SQUARED      :   .95309E+00


             MEAN OF THE RESIDUAL SERIES       :   .10804E-03
             STANDARD DEVIATION                :   .32153E+00
             NUMBER OF OBSERVATIONS            :   166
             MEAN DIVIDED BY THE STANDARD
             ERROR OF THE MEAN                 :   .43293E-02
```

THE AUTOCORRELATIONS

```
LAGS   1- 8     -.089    .014    .124    .071   -.031   -.039    .055    .014
STANDARD ERROR  (.078)  (.078)  (.078)  (.079)  (.080)  (.080)  (.080)  (.080)

LAGS   9- 16    -.098   -.021    .004   -.050   -.120    .035    .007   -.066
STANDARD ERROR  (.080)  (.081)  (.081)  (.081)  (.081)  (.082)  (.082)  (.082)

LAGS  17- 24     .003    .162    .000   -.114    .047   -.001   -.036   -.086
STANDARD ERROR  (.083)  (.083)  (.085)  (.085)  (.085)  (.086)  (.086)  (.086)

LAGS  25- 32     .133    .056   -.089   -.069   -.042   -.118   -.081   -.037
STANDARD ERROR  (.086)  (.087)  (.088)  (.088)  (.089)  (.089)  (.090)  (.090)

LAGS  33- 40    -.010    .024   -.005    .017   -.017    .043    .044   -.006
STANDARD ERROR  (.090)  (.090)  (.090)  (.090)  (.090)  (.090)  (.090)  (.090)

LAGS  41- 48     .084   -.112    .183   -.005   -.043   -.009   -.060   -.050
STANDARD ERROR  (.090)  (.091)  (.092)  (.094)  (.094)  (.094)  (.094)  (.094)
```

.
.
.

THE PARTIAL AUTOCORRELATIONS

LAGS 1- 8	-.089	.006	.127	.096	-.020	-.066	.025	.026
STANDARD ERROR	(.078)	(.078)	(.078)	(.078)	(.078)	(.078)	(.078)	(.078)

LAGS 9- 16	-.081	-.046	-.013	-.028	-.105	.016	.019	-.026
STANDARD ERROR	(.078)	(.078)	(.078)	(.078)	(.078)	(.078)	(.078)	(.078)

LAGS 17- 24	.006	.153	.035	-.114	-.027	-.043	-.010	-.024
STANDARD ERROR	(.078)	(.078)	(.078)	(.078)	(.078)	(.078)	(.078)	(.078)

LAGS 25- 32	.101	.079	-.030	-.109	-.122	-.125	-.037	-.029
STANDARD ERROR	(.078)	(.078)	(.078)	(.078)	(.078)	(.078)	(.078)	(.078)

LAGS 33- 40	-.016	.097	.042	-.023	-.064	.074	.049	-.040
STANDARD ERROR	(.078)	(.078)	(.078)	(.078)	(.078)	(.078)	(.078)	(.078)

LAGS 41- 48	.064	-.136	.097	-.040	-.011	-.015	-.044	-.035
STANDARD ERROR	(.078)	(.078)	(.078)	(.078)	(.078)	(.078)	(.078)	(.078)

THE RESIDUAL CROSS-CORRELATION ANALYSIS

INPUT SERIES : PREWHITENED OUT.DAT
OUTPUT SERIES : THE ESTIMATED RESIDUALS FROM THE TRANSFER FUNCTION MODEL

MEAN OF THE INPUT SERIES	:	.35922E-01
STANDARD DEVIATION	:	.14311E+01
NUMBER OF OBSERVATIONS	:	142

THE CROSS-CORRELATIONS

LAGS 0- 7	-.035	.023	.068	-.016	.025	.210	-.057	-.012
STANDARD ERROR	(.085)	(.085)	(.085)	(.085)	(.086)	(.086)	(.086)	(.087)

LAGS 8- 15	.080	-.025	-.038	-.075	.148	-.049	.013	-.008
STANDARD ERROR	(.087)	(.087)	(.088)	(.088)	(.088)	(.089)	(.089)	(.089)

LAGS 16- 23	-.107	.021	.021	-.052	-.005	.006	.026	.095
STANDARD ERROR	(.090)	(.090)	(.091)	(.091)	(.091)	(.092)	(.092)	(.092)

LAGS 24- 31	-.046	-.132	.023	.049	.108	.002	.015	.072
STANDARD ERROR	(.093)	(.093)	(.094)	(.094)	(.094)	(.095)	(.095)	(.096)

LAGS 32- 39	.091	-.082	-.134	.013	-.061	-.028	-.028	-.021
STANDARD ERROR	(.096)	(.097)	(.097)	(.098)	(.098)	(.099)	(.099)	(.100)

LAGS 40- 47	.015	-.084	.011	.040	-.047	-.042	.073	-.045
STANDARD ERROR	(.100)	(.101)	(.101)	(.102)	(.102)	(.103)	(.103)	(.104)

LAG 48	-.054
STANDARD ERROR	(.104)

INPUT SERIES : THE ESTIMATED RESIDUALS FROM THE TRANSFER FUNCTION MODEL
OUTPUT SERIES : PREWHITENED OUT.DAT

THE CROSS-CORRELATIONS

LAGS 0- 7	-.035	-.075	.011	.082	.052	-.044	-.002	.028
STANDARD ERROR	(.085)	(.085)	(.085)	(.085)	(.086)	(.086)	(.086)	(.087)

LAGS 8- 15	.041	-.013	-.028	-.022	-.049	.071	.016	-.094
STANDARD ERROR	(.087)	(.087)	(.088)	(.088)	(.088)	(.089)	(.089)	(.089)

LAGS 16- 23	-.067	.006	-.068	-.036	-.012	.039	.063	-.228
STANDARD ERROR	(.090)	(.090)	(.091)	(.091)	(.091)	(.092)	(.092)	(.092)

LAGS 24- 31	.104	.137	.012	-.001	-.068	.112	.011	-.070
STANDARD ERROR	(.093)	(.093)	(.094)	(.094)	(.094)	(.095)	(.095)	(.096)

PART C
AI Approaches to Learning from Data

9 Learning Tasks Studied in Artificial Intelligence

Robert C. Holte and Alan J. MacDonald

Department of Electrical Engineering and Electronics, Brunel University, England

1. A Framework for Describing Learning Systems

Most learning systems that are studied in Artificial Intelligence (AI) are engaged in two distinct tasks: a performance task and a learning task. The input/output behaviour required of the system defines the **performance task**. A simple performance task would be to find the largest integer in a given list of integers. A more complex performance task would be to prove that a given formula in Euclidean geometry was a theorem (relative to a given collection of axioms). An **instance** of the performance task is a complete set of specific inputs, e.g. an instance of the maximum integer task is a specific list of integers, and an instance of the theorem-proving task is a specific formula and collection of axioms.

The **learning task** is to improve the system's performance at the performance task. Commonly used measures of improvement are:
  (1) the increase in the number of instances for which the system produces an output;
  (2) the increase in the number of instances for which the output produced by the system is correct;
  (3) the increase in the efficiency of the system.

Most learning systems alternate between these two tasks, typically processing one instance of the performance task to completion and then doing some processing associated with the learning task before processing the next instance of the performance task. Because of this clear separation of the two tasks, it is natural to describe a learning system in terms of two components: a **learning component** which never changes, and a **performance component** which is replaced or modified as deemed appropriate by the learning component.

To describe the relationship between the two components more precisely it is useful to describe the performance component in terms of two parts: a part which is supplied by the learning component, called the **declarative aspect**, and a part which utilises the declarative aspect in carrying out the performance task, called the **interpreter**. For example, in the ID3 system [24] the declarative aspect of the performance component is a decision tree, and the interpreter is a procedure which uses a decision tree to classify objects. As a second example, if the performance task involved interpolating a function of N variables, the declarative aspect might be a vector of real numbers, and the interpreter a procedure which uses the vector of numbers as coefficients in defining a function which serves as the basis for interpolation. Only the declarative aspect of the performance system (e.g. the decision tree, or vector of coefficients) can change; the interpreter cannot.

Two types of information are available to the learning component to guide its search. The first type is a record of the performance associated with each declarative aspect that has been proposed (i.e. forwarded to the interpreter), including the performance task instances to which each was applied and the output that was produced. This record may also include a trace of the steps followed by the interpreter in applying a given declarative aspect to a given performance task instance.

Performance assessment is a second type of information that is often available to the learning component. It is an indication of the relation between the performance achieved by each of the proposed declarative aspects and an independently defined **target performance.** For example, the learning component is often supplied with the correct output for a specific instance of the performance task. By accumulating this information, the learning component is able to build up an increasingly accurate characterisation of the target performance. When this type of information is available the learning task is to find a declarative aspect which results in performances that are as close to the target as possible.

Most of the learning systems studied in AI are designed to perform satisfactorily only if the information presented to both the learning and performance components is correct. For example, if the performance task was to classify integers as even or odd, after being presented with the incorrect fact that twelve is odd, the learning component would restrict its search to those declarative aspects which resulted in twelve being classified as odd; never would the correctness of the fact be questioned. If a contradictory fact were subsequently presented (e.g. "twelve is even"), most systems would be unable to take corrective action.

2. Learning a Classification Procedure Given Classified Examples

When presented with a member of a prespecified universe of objects, a **classification procedure** selects a subset of a small prespecified set of possible class names.[1] A classification procedure is **partial** if there exists an object for which it selects no class name, and **total** if there does not. In the learning task most commonly studied in AI, called **learning from examples**[8] or **concept learning**[12], the performance component is a classification procedure, and the learning component is provided with a set of positive examples, where a **positive example** is an object together with its correct classification. In most systems, the learning component is also provided with set of negative examples, where a **negative example** is an object together with the name of a class of which it is not a member (i.e. an incorrect classification).[2]

The learning task is to find a declarative aspect which, when interpreted, correctly classifies all objects in the universe. In many learning components this search is guided by heuristics that are likely to lead to compact declarative aspects, but in this learning task the efficiency of the performance component is of secondary importance.

---

[1] a selected subset containing more than one class name might indicate (1) the object is a member of all the classes named, or (2) the object is a member of at least one of the classes named. Many classification procedures do not produce subsets containing more than one class name.

[2] if classes are known to be mutually exclusive, some information normally provided by negative examples can be inferred from positive examples

Every learning system requires a language in which to represent objects.[3] A language for representing triangles with sides of specific length might be built out of the following primitives

```
ANGLE1 = a
ANGLE2 = b
ANGLE3 = c
SIDE1 = x
SIDE2 = y
SIDE3 = z
```

and the logical connective "&" ("and"), where a, b,c (measured in degrees, say) and x, y, z are appropriate constants. This type of language is called an **attribute-value** language. The attributes (ANGLE1, ANGLE2, ANGLE3, SIDE1,SIDE2, SIDE3) denote particular properties of an object (i.e. ANGLE1 refers to some specific angle in a triangle. Thus, in this language there are 6 different descriptions of any triangle. This is not a desirable feature of a representation language.).

In an attribute-value language objects are represented by a fixed number of properties. There are many universes for which no attribute-value language can adequately describe every object in the universe. For example, the triangle representation language can easily be extended to describe quadrilaterals, or any other polygons with a specific number of sides. But it cannot easily be extended to describe general polygons where the number of sides is not known (or at least bounded) in advance.

If every object in a given universe can be represented in an attribute-value language, the objects are said to be **atomic.** A **compound object** is a structured collection of atomic objects, that is, a set of atomic objects among which certain relations hold. For example, a collection of points, certain of which are connected by line segments, is a compound object; each point is an atomic object and the pattern of connectedness is a relation that holds between atomic objects.

Languages for representing compound objects are called **relational languages.** If the same atomic object may participate in several relations it must be possible to refer to specific atomic objects. Typically, relational languages provide a facility for naming atomic objects; the attributes of atomic objects are then represented as functions of these names. Some relational languages allow compound objects to be represented as structured collections of atomic objects and other compound objects, and support this ability to describe compounds objects hierarchically by providing a facility for naming compound objects and relations which take compound objects as parameters. Continuing the example of the previous paragraph, it is convenient to be able to name specific line segments, and the angle between two specific line segments that share an endpoint. A relational language that can represent all such geometric figures can be built out of the following primitives:

---

[3] The language for representing objects is the same in both the learning and performance components.

A, B, C...   -- unique names for the points in the figure

CONNECTED(p1,p2) -- asserts that there exists a line segment
                    joining points p1 and p2.

SEGMENT(p1,p2)   -- identifies the line segment joining points p1
                    and p2. By convention, a reference to
                    SEGMENT(p1,p2) implies CONNECTED(p1,p2).
ANGLE(p1,p2,p3)   -- identifies the angle formed by the line segment
                    joining points p2 and p1 and the line segment
                    joining points p2 and p3. By convention,
                    a reference to ANGLE(p1,p2,p3) implies
                    CONNECTED(p2,p1) and CONNECTED(p2,p3).

SEGMENT(p1,p2) ≅ SEGMENT(p3,p4) -- asserts that the two line
                                     segments are congruent
ANGLE(p1,p2,p3) ≅ ANGLE(p4,p5,p6) -- asserts that the two
                                     angles are congruent

p1 IS_ON SEGMENT(p2,p3) -- asserts that point p1 is on the line
                             segment joining points p2 and p3.

    & -- logical conjunction

In this language, the object in figure 1(a) could be represented as

        SEGMENT(A,B) ≅ SEGMENT(B,C)
     & CONNECTED(A,C)

and the object in figure 1(b) as

        SEGMENT(W,X) ≅ SEGMENT(X,Y)
      & CONNECTED(W,Y)
      & CONNECTED(W,Z)
      & Y IS_ON SEGMENT(X,Z)

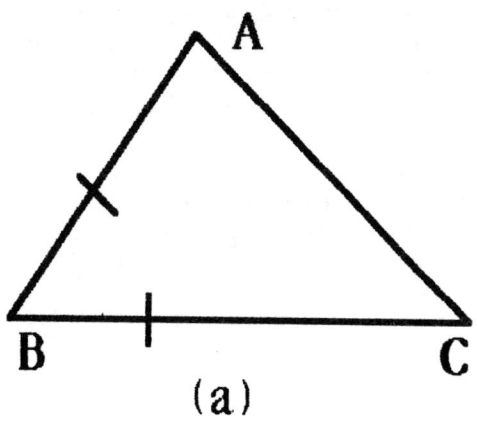
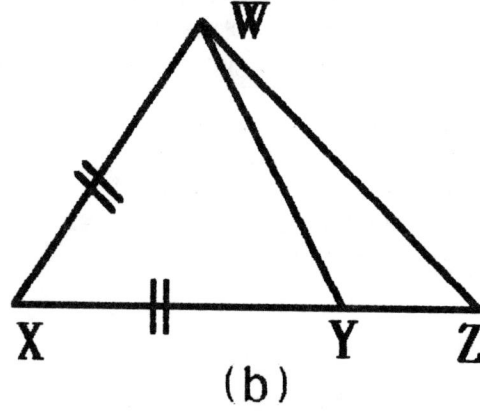

Figure 1

Learning systems also require a language in which to represent declarative aspects. The expressiveness of this language has a direct bearing on the difficulty of the learning task. As a general rule, the more expressive this language, the more difficult the learning task. A declarative aspect for a classification procedure is typically represented as a set of rules of the form:[4]

```
CLASSIFY object
    AS class name
    IF conditions
```

where conditions is an expression which may or may not be satisfied by a given object. If an object satisfies a rule's conditions, the rule's class name is used as the classification of the object. The language in which the conditions are represented is often very similar to the language for representing objects. If conditions were expressed in the attribute-value language for triangles, the declarative aspect for the class of right angle triangles would require three rules:

```
CLASSIFY object
    AS  "right_angle"
    IF  ANGLE1 = 90

CLASSIFY object
    AS  "right_angle"
    IF  ANGLE2 = 90

CLASSIFY object
    AS  "right_angle"
    IF  ANGLE3 = 90
```

The attributes in the language for representing objects are chosen specifically to describe properties of individual objects. But the conditions in a rule describe a set of objects (namely, the set of objects that satisfy the conditions), and frequently it is necessary to augment the language for representing objects in order to be able to represent all the declarative aspects of interest. For example, no declarative aspect corresponding to the set of isosceles triangles (definition: SIDE1=SIDE2 or SIDE1=SIDE3 or SIDE2=SIDE3) can be represented in the triangle representation language. Likewise, no declarative aspect corresponding to the set of acute triangles (definition: SIDE1 < 90 & SIDE2 < 90 & SIDE3 < 90) can be represented in this language.

Representing the conditions of a rule with a relational language requires a minor change in the interpretation of the names of atomic objects. When the language is used to represent compound objects, these names denote specific atomic objects and are interpreted as constants. When it is used to represent conditions in a rule, these names are interpreted as existentially quantified variables. Using this type of language, a declarative aspect will be represented by a set of rules of the form

```
CLASSIFY compound object
    AS class name
    IF THERE EXISTS var1,var2,...      (a finite number of variables)
    SUCH THAT  conditions
```

---

[4] decision trees are a variant of this basic scheme

The intuitive meaning of a rule of this type is that a given compound object should be classified using the rule´s class name if any part of the compound object satisfies all the conditions. Thus, providing some three points in the given compound object form an isosceles triangle, the rule

```
CLASSIFY compound object
  AS "contains an isosceles triangle"
  IF THERE EXISTS var1, var2, var3
  SUCH THAT
      SEGMENT(var1,var2) ≅ SEGMENT(var1,var3)
    & CONNECTED(var2,var3)
```

will classify the compound object as "contains an isosceles triangle".

Determining if a compound object satisfies a rule´s conditions is now a more complicated process. First, each variable in the rule is **bound** to some atomic object in the compound object. The result of binding var1 to obj1 is that all occurrences of var1 in the conditions are replaced by obj1. Any variable may be bound to any atomic object, and therefore it is possible for two or more variables to be bound to the same atomic object. The rule for determining if a compound object contains an isosceles triangle is incorrect because of this: the conditions are satisfied when all three variables are bound to the same atomic object. To express the requirement that two specific variables must not be bound to the same atomic object, a predicate such as DISTINCT(var1,var2) is usually added to the language.

When substitutions have been made for all the variables, the conditions will be a logical expression describing relations between the atomic objects in the given compound object. The compound object **satisfies** the conditions if this logical expression is a logical consequence of the description of the compound object and any **background axioms** that are available. Three of the background axioms in the geometry domain are

```
(AX1)    CONNECTED(p1,p1)
(AX2)    CONNECTED(p1,p2)  =>  p2 IS_ON SEGMENT(p1,p2)
(AX3)    p1 IS_ON SEGMENT(p2,p3)  =>  CONNECTED(p1,p2) & CONNECTED(p1,p3)
```

Consider the following rule.

```
CLASSIFY  compound object
  AS "like_figure_1"
  IF THERE EXISTS  var1, var2, var3, var4
  SUCH THAT

      SEGMENT(var1,var2)  ≅ SEGMENT(var2,var3)
  &   CONNECTED(var1,var3)
  &   CONNECTED(var1,var4)
  &   var3 IS_ON SEGMENT(var2,var4)
  &   CONNECTED(var3,var4)

  &   DISTINCT(var1,var2) & DISTINCT(var1,var3) & DISTINCT(var1,var4)
  &   DISTINCT(var2,var3) & DISTINCT(var2,var4)
```

There are many ways in which the variables in this rule may be bound to the atomic objects in the compound object in figure 1(a). Binding var1 to A, var2 to B, and var3 and var4 to C, the conditions in this rule become[5]

---

[5] ignoring the occurrences of DISTINCT, which are all true for both this set of bindings and the next

```
        SEGMENT(A,B) ≅ SEGMENT(B,C)
    & CONNECTED(A,C)
    & CONNECTED(A,C)
    & C IS_ON SEGMENT(B,C)
    & CONNECTED(C,C)
```

The first three conjuncts occur in, and are therefore logical consequences of, the description of the object in figure 1(a). The fourth is a consequence of axiom AX2 and the fifth is a consequence of axiom AX1. Therefore, the compound object in figure 1(a) would be classified by this rule as "like_figure_1".

The compound object in figure 1(b) would also be classified by this rule as "like_figure_1". Binding var1 to W, var2 to X, var3 to Y, and var4 to Z, the conditions in this rule become

```
        SEGMENT(W,X) ≅ SEGMENT(X,Y)
    & CONNECTED(W,Y)
    & CONNECTED(W,Z)
    & Y IS_ON SEGMENT(X,Z)
    & CONNECTED(Y,Z)
```

The first four conjuncts are logical consequences of the description of the object in figure 1(b). The final conjunct, CONNECTED(Y,Z), is a consequence of Y IS_ON SEGMENT(X,Z) and axiom AX3.

The Conventional Limits of this Learning Task

Conventional use of the term "learning (classification procedures) from examples" excludes some of the tasks that are encompassed by the definition given at the beginning of this section.

According to convention, "learning from examples" does not include the learning of classification procedures in which the classification of an object depends on either

(a) other classifications of the object[6]; or
(b) the classifications of its parts.

Both types of dependency are illustrated in the following declarative aspect for a procedure that classifies a given integer list as "sorted".

```
    CLASSIFY  integer_list
        AS  "short"
        IF  LENGTH integer_list  <  2

    CLASSIFY  integer_list
        AS  "sorted"
        IF  "short" = CLASSIFY integer_list
    CLASSIFY integer_list
        AS  "sorted"
        IF  FIRST_ELEMENT  <  SECOND_ELEMENT
          & "sorted" = CLASSIFY (integer_list WITHOUT  FIRST_ELEMENT)
```

Rule sets with unrestricted dependencies between rules are computationally universal. The task of learning an arbitrary computation from input-output examples has been the subject of considerable theoretical [3] and

---

[6] recent exceptions which study the learning of dependent classes from examples are [26] and [10].

practical[7] investigation quite separate from the investigation of the
conventional "learning from examples" task.

Only one type of dependency is permitted by convention, and that is that
the classes may be stipulated to be mutually exclusive. Various means can
be employed to enforce mutual exclusivity without including in the rule
language a mechanism that allows rules to make explicit reference to each
other's conclusions.

Also according to convention, "learning from examples" requires class names
to be constants. Therefore it excludes classification procedures in which
the correct class name of an object is constructed from the values of
attributes or subobjects of the object. Suppose objects are strings of
alphabetic characters, and class names are single alphabetic characters.
Given the information that "sword" is in class "s" and "award" is in class
"a", a learning component might produce a declarative aspect of a procedure
that classifies an object on the basis of its first letter (FIRST_LETTER is
assumed to be an attribute of the object).

```
CLASSIFY  object
   AS  var1
   IF THERE EXISTS var1
   SUCH THAT
        FIRST_LETTER  =  var1
```

The class name produced when the conditions of this rule are satisfied is
constructed from the value of the FIRST_LETTER attribute. Thus, this one
rule may produce different class names, depending on the object it is
classifying. Also, the class names that can be produced by this rule are
not restricted to any prespecified set, nor to the set of class names that
have accompanied the classified examples; given "zebra", this rule would
produce class name "z" without having been given any classified objects in
class "z".

As a second example, the following is a declarative aspect of a procedure
which identifies two triangles in a given compound object that are
congruent as a consequence of the "side-angle-side" (SAS) postulate.

```
CLASSIFY  object
   AS "contains congruent triangles (vA,vB,vC) and (vX,vY,vZ)"
   IF THERE EXISTS  vA, vB, vC, vX, vY, vZ
   SUCH THAT
        ANGLE(vA,vB,vC)  ≅  ANGLE(vX,vY,vZ)
      & SEGMENT(vA,vB)   ≅  SEGMENT(vX,vY)
      & SEGMENT(vB,vC)   ≅  SEGMENT(vY,vZ)
      & DISTINCT(vA,vB) & DISTINCT(vA,vC) & DISTINCT(vB,vC)
      & DISTINCT(vX,vY) & DISTINCT(vX,vZ) & DISTINCT(vY,vZ)
```

Tasks that involve learning classification procedures of this kind will now
be examined.

---

[7] [30] reviews systems for synthesising LISP programs. [27] describes a
system for synthesising PROLOG programs which has been adapted to other
programming languages in [13].

3. Improving Performance at a Problem-Solving Task Given Worked Problems

Given:

(1) a set of possible **problems**;

(2) a **criterion** that defines the subset of possible problems which are to be considered **solved**; and

(3) a set of **operators** that transform one problem into another;

the **problem-solving task** is to transform any given problem into one of the solved problems.

A system for a particular problem-solving task requires definitions of the operators and a language in which to represent problems. The definition of an operator specifies the transformation that the operator performs and the conditions under which the problem-solving system is to consider applying the operator. Typically, operator definitions are represented in a problem-solving system as

```
OPERATOR  name
PARAMETERS var1, var2,...      (variables used in the definition)
TRANSFORM  given_problem
     INTO  output_problem
IF  condition
```

For example, in the geometry domain of the previous section, problems would be geometric figures, represented as a set of assertions. Operators would be rules of inference, such as the SAS postulate or the background axioms, which transform a set of assertions by adding new assertions to it. Represented as this type of operator, the SAS postulate would be:

```
OPERATOR  "SAS"
PARAMETERS  vA, vB, vC, vX, vY, vZ
TRANSFORM  assertions_about_figure
     INTO  (ADD_ASSERTION  TRIANGLE(vA,vB,vC) ≅ TRIANGLE(vX,vY,vZ)
            TO   assertions_about_figure)
IF
    ANGLE(vA,vB,vC)  ≅ ANGLE(vX,vY,vZ)
  & SEGMENT(vA,vB)   ≅ SEGMENT(vX,vY)
  & SEGMENT(vB,vC)   ≅ SEGMENT(vY,vZ)
  & DISTINCT(vA,vB) & DISTINCT(vA,vC) & DISTINCT(vB,vC)
  & DISTINCT(vX,vY) & DISTINCT(vX,vZ) & DISTINCT(vY,vZ)
```

A problem-solving system is given a problem as input and produces three distinct products[8]

(a) a sequence of operators (with specific values for their parameters), called a **solution** of the problem, that transforms the given problem into a solved problem;

(b) the knowledge that a solution does or does not exist for the given problem;

(c) the solved problem into which the solution maps the given problem.

Different problem-solving tasks may consider any one of these to be the primary product, the others being mere by-products.

(a) In a planning task, the problems are "states", the operators are the possible actions (which modify the existing state), and the solved problems are the desired states. The primary product is the plan, i.e. the sequence of actions that transforms the given state into a desired state.

---

[8] of course (a) and (c) are produced only if they exist

(b) In a theorem-proving task, the problems are logical formulae, the operators are the rules of inference, and the solved problems are the axioms. The primary product is the knowledge that the given formula is or is not a theorem.

(c) In a symbolic algebra task, such as symbolic integration, the problems are algebraic expressions, the operators are the rules of algebraic manipulation, and the solved problems are members of a special class of formula (e.g. in the integration domain, formula containing no integration sign). The primary product is the formula in the special class into which the given formula can be transformed.

The task of improving a problem-solving system is a learning task. AI has studied two learning tasks of this type, each corresponding to a different notion of improvement.

3.1. Improving the Number of Problems Solved Correctly

In the first, the problem-solving system is partial and the learning task is to increase the number of problems that the problem-solving system can solve correctly. The input to the learning component is a set of problem-solving **traces**, where a trace is a sequence of problems that terminates with a solved problem. It is assumed that each problem in the sequence is the result of applying a single operator to its predecessor.

The usual approach to this learning task [9, 22, 28, 29, 31] is to have the learning component create a new operator for each step (pair of consecutive problems) in the trace which it cannot recognise as an application of one of the existing operators. It attempts to create operators that are applicable to a wide range of problems. In its most general form this learning task is called program synthesis from computation traces[4].

3.2. Improving Problem-Solving Efficiency

In the second and more frequently studied learning task of this type, it is the efficiency of the problem-solving system which the learning component seeks to improve. The problem-solving system is assumed to be total and correct in principle, although in practice it may be unable to solve some problems correctly because of its inefficiency. The input to the learning component is a set of **worked examples**, where a worked example consists of a problem and its solution.

Typically, the problem-solving system employs a simple breadth-first forward search to solve each problem. If the **distance** from one problem to another is defined to be the length of the shortest operator sequence that transforms the first into the second, breadth-first search is guaranteed to find the solved problems that are nearest to the initial problem. One measure of the **efficiency** of a problem-solving system is the total number of problems that are inspected before a solved problem is found; the smaller this number, the more efficient the problem-solving system. If the nearest solved problem is at distance d, and the average number of operators that the problem-solving system considers applying to a typical problem (called the **branching factor**) is B, then the efficiency of a breadth-first search is B^d.

Therefore there are two means by which to improve the efficiency of a problem-solving system which uses breadth-first search:
  (1) reduce the branching factor -- that is, make the problem-solving system more selective in applying operators to each problem; and
  (2) reduce the (average) distance to the nearest solved problem. For example by promoting a sequence of operators to the status of a single operator (called a **macro-operator**) the distance between problems connected by this sequence is reduced to 1.

Learning When to Apply Individual Operators

To make the problem-solving system more selective in its use of an operator the learning component simply adds conditions to those already present in the operator´s definition; it does not change the definition of the operator in any other way. In particular it does not add to or modify the variables that appear in the transformation function in the "INTO" field. If the operator name is regarded as a class name, and the problems that satisfy the conditions are regarded as members of the operator´s class, this learning task seems to satisfy the definition of the task of learning classification procedures discussed in the section 2. Indeed most of the learning systems for the present task are straightforward adaptations of systems for the previous task.

There is one important difference between the two tasks: the nature of the input to the learning component. In the earlier task, the learning component was given classified examples, which stated with certainty that a particular object was a member of one particular class (positive information) or that the object was definitely not a member of particular classes (negative information). The worked examples that are given to the learning component in the present learning task readily provide an exact counterpart of the positive information; if operator OP is applied to problem P in a worked example then it is certain that P is a member of the class associated with OP. Positive information establishes a limit on the selectivity of the problem-solving system in the sense that no problem-solving system should be so selective that it fails to apply an operator to a problem when that operator has been applied to that problem in one of the worked examples.

It is negative information that is needed to increase the selectivity of the problem-solving system, but reliable negative information is not readily available in the present task. In most domains, the classes associated with different operators are not mutually exclusive; for example, there may be multiple solutions to some problems. The LEX system [21] invests considerable additional search in ensuring that the solution presented in each worked example is unique. In the PLS system [25] the problem-solving system is based on best-first search rather than breadth-first search, thereby gaining the flexibility to apply operators selectively without relinquishing the possibility of applying any operator to any problem.

Adapting Previous Solutions to New Problems

The intuition behind this learning task is that if a sequence of operators transforms one problem into a solved problem, it may do likewise to other "similar" problems. There are two formulations of this learning task.

In the first formulation, the learning task is to derive from a worked example an operator that is likely to reduce the distance to a solved problem from a fairly large number of problems. Since the derived operator is typically a sequence of existing operators it is called a macro-operator, and this formulation is called **learning macro-operators.** There are two main subtasks within this task:

(1) creating a transformation function for the macro-operator; and
(2) determining the conditions under which to apply the macro-operator.

In its full generality, the task of learning macro-operators has much in common with the task of learning new operators (section 3.1). The major difference is that systems which learn new operators do not have any information corresponding to the detailed information about the sequence of operators that appeared in the worked example.

Most of the present research in this area [7, 14] is much more concerned with the second subtask than the first. The transformation function of the macro-operator is created by generalising, in a prespecified and somewhat mundane way, the transformation function of the sequence of operators in the worked example (e.g. certain constants are replaced by variables). The conditions for the macro-operator are derived from properties of the problem in the worked example and the conditions in the definitions of the operators in the sequence. Often the conditions that are derived are more selective than is strictly necessary.

In the second formulation of this learning task, it is the performance component which adapts previous solutions to new problems. The learning component supplies it with slightly modified versions of the worked examples, called **prototypes** (or exemplars). Thus each prototype consists of a problem and a solution. A special interpretive process, often called **analogical reasoning**, is used to match each prototype problem to a given problem. If the match succeeds, it is used to guide the adaptation of the prototype solution. If the adapted prototype solution does not constitute a solution to the given problem, it may be used as a starting point for the search for a solution.

Contrary to the framework given in section 1, in systems that learn by analogical reasoning [1, 5, 23] the learning task is not entirely performed by the learning component. Comparing the second formulation of the present learning task with the first, both the major subtasks of macro-operator learning are performed by the analogical reasoning process. It determines the conditions on a macro-operator in the sense that it determines whether a given problem matches any of the worked examples. And it creates the transformation function a macro-operator in the sense that it derives a solution to the given problem from the prototype solutions. [2] observes that macro-operators can be created by "compiling" the analogical reasoning process.

However formulated, this learning task aims to increase the efficiency of the problem-solving component by supplementing the original set of operators with "shortcuts", i.e. macro-operators or prototypes, that often lead directly to solutions. Each added shortcut increases the branching factor. Therefore, unless the new shortcuts result in considerably shorter solutions the efficiency of the problem-solving system will decrease rather than increase. [20] reports an experiment measuring the degree of this effect.

4. Discovery Systems

To categorise a learning system as a discovery system is to place it in a class which is somewhat inhomogeneous and consequently difficult to characterise precisely. In general terms the task of a discovery system is to organise or impose structure on a set of instances (objects or events). This organisation may take different forms. One form of organisation is a taxonomy of objects; another form is a set of laws describing relationships between objects.

In a typical discovery task the organisation of the given objects or events is an end in itself. That is, a discovery task is often defined without reference to a performance component that exploits the organisation produced by the discovery component. When a performance task is associated with a discovery task, the quality of performance assessment available to the discovery component is usually very poor compared to that available in the tasks previously discussed.

4.1. Conceptual Clustering

The conceptual clustering [19] task can be stated as follows. Given a set of examples (object or event descriptions), produce a classification procedure. A **clustering** is the set of classes, or **clusters**, determined by the classification procedure. Unlike the concept learning task (section 2), in the conceptual clustering task the learning component receives no information about the correct classification of any example. A second difference is that the clustering produced is generally required to be hierarchical. That is, each cluster may itself be subdivided into "sub-clusters" which in turn may be divided into sub-sub-clusters and so on. Clusters are often required to be disjoint.

The sole means by which the learning component discriminates between different possible declarative aspects is by applying a heuristic **preference criterion**. Such a criterion defines an order (possibly partial) on the set of declarative aspects. The preference criterion may be based on: purely syntactic properties of the declarative aspect; on properties of the clustering obtained; or on a combination of the two. A **syntactic property** is one whose definition depends on the language in which declarative aspects are represented. For example, shorter declarative aspects are often preferred over longer ones, but the length of a declarative aspect is entirely determined by the language in which it is represented. A commonly used property of the clustering obtained is a measure of the disjointness of classes, or a measure of the distance between classes[9].

The learning task, therefore, is to find a declarative aspect which is optimal according to the preference criterion. The following example illustrates how a preference criterion influences the clustering obtained. Suppose four objects corresponding to the following descriptions are to be clustered:

---

[9] such as those used in numerical clustering. For a comparison of conceptual clustering and the agglomerative algorithm often employed in numerical classification procedures see [6].

```
COLOUR = RED    & SHAPE = SQUARE
COLOUR = BROWN  & SHAPE = TRIANGLE
COLOUR = BLUE   & SHAPE = SQUARE
COLOUR = GREEN  & SHAPE = TRIANGLE
```

The preference criterion, P, is a function of a clustering, z, of the form

$$P(z) = c*C(z) + m*M(z)$$

where $C(z)$ is the complexity of the declarative aspect of z (defined below), $M(z)$ is a measure of the "mismatch" between the clustering z and the data points (defined below), and c and m are constants which can be chosen to reflect the relative undesirability of complexity and mismatch. Clusterings with lower values of P are to be preferred.

Define the distance between attribute values to be

```
d(SQUARE,TRIANGLE)              = 1
d(RED,BROWN) = d(BLUE,GREEN)    = 1
d(RED,GREEN) = d(BROWN,BLUE)    = 2
d(RED,BLUE)  = d(BROWN,GREEN)   = 3
```

and the distance between two objects to be the sum of the distances between corresponding attribute values. The mismatch, $M(z)$, is the ratio of mean intra-cluster distance to mean inter-cluster distance.

The declarative aspect of a clustering consists of a set of rules expressed in an attribute-value language, with one rule for each cluster. The language includes the symbol "|" signifying logical or. A term is an expression of the form ATTRIBUTE = VALUE. The complexity of a clustering, $C(z)$, is the mean number of terms in the declarative aspects of the clusters in z.

If 0.5 and 3 are chosen for c and m respectively, penalising mismatch more heavily than complexity, the following clustering has the lowest value of P.

```
CLASSIFY object AS CLASS 1
IF
   COLOUR = RED
 | COLOUR = BROWN

CLASSIFY object AS CLASS 2
IF
   COLOUR = BLUE
 | COLOUR = GREEN
```

If 2 and 0.6 are chosen for c and m respectively, penalising complexity more heavily than mismatch, the following clustering has the lowest value of P of any two class clustering.

```
CLASSIFY object AS CLASS 1
IF
   SHAPE = SQUARE

CLASSIFY object AS CLASS 2
IF
   SHAPE = TRIANGLE
```

In this example there are few enough clusterings to allow identification of the optimal clustering by exhaustive enumeration. Generally the set of possible clusterings is enormous and it is necessary to perform a heuristic search to find even a reasonably good clustering.

4.2. Inference of Empirical Laws

Given a set of observations, this task is to infer a set of general laws which are not refuted by the given observations[15]. A set of laws divides the set of possible observations into two classes: **prohibited observations** are those that refute the laws, and **permitted observations** are those that do not refute the laws. This learning task can therefore be related to the concept learning task (section 2). The laws and observations upon which they are based serve as the declarative aspect for a procedure that classifies observations as "permitted" or "prohibited". In this task, the only observations available to the learning component are those classified as "permitted" by the target set of laws.

An observation is an assertion about a particular property of an object, such as TASTE(NaCl,SALTY), or relation between objects, such as REACTS(HCl,NaOH,NaCl,H2O). In most discovery systems, objects can be uniquely identified. This enables several observations to refer to the same object. For example, "NaCl" refers to the same object in all observations in this section.

By themselves, observations indicate only what is possible, they do not indicate what is necessary. For example, TASTE(NaCl,SALTY) means "it is possible for the object NaCl to taste salty" and not "the object NaCl must always taste salty". Similarly, when an observation such as NOT TASTE(NaCl,SOUR) is available it means "it is possible for NaCl not to taste sour" and not "it is not possible for NaCl to taste sour".

A **law** expresses a relationship between observations. For example, the statement that "NaCl always tastes salty" is a relationship between separate observations of the taste of NaCl. The following law expresses this relationship (s1, s2, t1, and t2 are variables).

```
NAME:  "Law-1"
OBSERVATIONS:  TASTE(s1,t1), TASTE(s2,t2)
RELATIONSHIP:  s1=s2=NaCl => t1=t2
```

After a single observation that the taste of NaCl is salty, this law can be reformulated as

```
NAME:  "Law-2"
OBSERVATIONS:  TASTE(s1,t1)
RELATIONSHIP:  s1=NaCl => t1=SALTY
```

One law is said to be **more general** than another if the number of sets of observations which refute first law is greater than the number which refute the second. For example, consider the following law, which asserts that the taste of any substance is unique.

```
NAME:  "Law-3"
OBSERVATIONS:  TASTE(s1,t1), TASTE(s2,t2)
RELATIONSHIP:  s1=s2 => t1=t2
```

Any observations which refute Law-1 will also refute Law-3. There are also sets of possible observations, such as

```
TASTE(SUBSTANCE-X,SOUR)
TASTE(SUBSTANCE-X,BITTER)
```

(where SUBSTANCE-X is not NaCl) which refute Law-3 but not Law-1. Thus the number of sets of observations which refute Law-3 is greater than the number which refute Law-1, and so Law-3 is more general than Law-1.

A law of the form "P => Q" is refuted by a set of observations which satisfy P and contradict Q (i.e. satisfy NOT Q). Some laws of this form cannot be refuted because no set of available observations contradicts Q. For example,

```
    NAME:  "Law-4"
    OBSERVATIONS:
                    TASTE(s1,t1), TASTE(s2,t2), TASTE(s3,t3),
                    REACTS(s4,s5,s6,s7)
    RELATIONSHIP:
                    s1=s4     &   s2=s5      &   s3=s6
                &  t1=SOUR  &   t2=BITTER  &   t3=SALTY
                => s7=CO2
```

is not refuted by the observations

```
        TASTE(HCl,SOUR),  TASTE(NaOH,BITTER),  TASTE(NaCl,SALTY),
    and  REACTS(HCl,NaOH,NaCl,H2O)
```

because   REACTS(HCl,NaOH,NaCl,H2O)   does   not   contradict REACTS(HCl,NaOH,NaCl,CO2). Indeed, no observations which are permitted by the target laws can contradict REACTS(x,y,z,CO2), and therefore Law-4 cannot be refuted. In order to increase the number of laws which are refutable, discovery systems are often provided with an initial set of laws, such as Law-3 or the following law which states that there are unique reaction products for any pair of reactants.

```
    NAME:  "unique reaction products"
    OBSERVATIONS:  REACTS(x1,y1,p1,q1), REACTS(x2,y2,p2,q2)
    RELATIONSHIP:  (x1=x2  &  y1=y2) => (p1=p2  &  q1=q2)
```

The learning component which produces sets of laws is often engaged in several different discovery subtasks[16]. One discovery subtask is the formulation of laws from observations. If in all of the reactions that have been observed it has also been observed that the first reactant was sour, the second reactant bitter, the first reaction product salty, and the second reaction product water, a learning component might formulate

```
    NAME:  "reaction law 1"
    OBSERVATIONS:
                    TASTE(s1,t1), TASTE(s2,t2), TASTE(s3,t3),
                    REACTS(s4,s5,s6,s7)
    RELATIONSHIP:
                    t1=SOUR  &   t2=BITTER  &   t3=SALTY
                => s1=s4     &   s2=s5      &   s3=s6     &  s7=H2O
```

which states that all sour substances react with all bitter substances to produce all salty substances and water. A different law that might be formulated from the same observations is

```
    NAME:  "reaction law 2"
    OBSERVATIONS:
                    TASTE(s1,t1), TASTE(s2,t2), TASTE(s3,t3),
                    REACTS(s4,s5,s6,s7)
    RELATIONSHIP:
                    s7=H2O
                => s1=s4     &   s2=s5      &   s3=s6
                &  t1=SOUR  &   t2=BITTER  &   t3=SALTY
```

which states that whenever the second reaction product is water the reactants must taste sour and bitter and the first reaction product must taste salty.

Both these laws are too general, and will be refuted eventually. A second discovery subtask is to specialise (generalise) existing laws which are too general (specific). For example, reaction law 1 might be specialised to

```
NAME:  "reaction law 3"
OBSERVATIONS:
            TASTE(s1,t1), TASTE(s2,t2), TASTE(s3,t3),
            REACTS(s4,s5,s6,s7)
RELATIONSHIP:
            t1=SOUR  &  t2=BITTER  &  t3=SALTY
        &  s1=s4    &  s2=s5
       => s3=s6     &  t3=SALTY    &  s7=H2O
```

which states that if a sour substance reacts with a bitter substance the reaction products will be a salty substance and water. This law is less general than reaction law 1 because fewer sets of observations will refute it.

A third discovery subtask is defining new classes or properties of objects. For example, the learning component might find it useful to formulate the following definition.

```
ALKALINE(x)       (x any substance)
IF  TASTE(x,BITTER)  OR  FEEL(x,SOAPY)
```

A generalisation of reaction law 3 can be created by replacing the premise t2=BITTER with ALKALINE(s2).

A fourth discovery subtask is conjecturing properties of the relations and attributes which occur in observations. One very important property of an attribute (or relation) $P(x,y)$ is that for each value of argument x there is a unique value of argument y such that $P(x,y)$ holds. Law-1 expresses this property of the TASTE attribute, and the unique reaction products law expresses a similar property for the REACTS relation. Other properties that might be conjectured are the transitivity or commutativity of a relation.

The difficulty of the empirical law formation task is largely a function of the number and diversity of laws (relationships between observations) that might be required to capture the regularities in a set of observations. For example, the task of discovering numerical laws is straightforward if the laws are known to be in a family of easily distinguishable functions (e.g. polynomials of degree N). To obtain reasonable performance when the set of possible laws is large and diverse, it is necessary to provide a discovery system with correspondingly powerful heuristic knowledge. Typical "knowledge intensive" discovery systems are discussed in the following section.

4.3. The General Discovery Task

Because the general task of discovery is open ended, systems for this task must be open ended. [11, 17, 18] describe a general architecture for open ended discovery systems. In this architecture, all attributes, relations, objects, and functions are referred to as **concepts**. Statements about concepts are called **conjectures**. Given an initial set of concepts, a discovery system formulates new concepts and conjectures, and continuously ranks them according to their **worth**, or **interestingness**. A concept or conjecture which has been formulated is said to be **discovered** only if it is highly ranked by the system for a sustained period. The global objective of the system is to discover highly interesting collections of concepts and conjectures.

Heuristics are used to evaluate the worth of a concept or conjecture. Some of these will be domain specific, i.e. will apply only when specific concepts are included in the initial set of concepts, but many will be of general applicability. One general heuristic is "a concept is interesting if it can be constructed in a number of independent ways". Another is "a concept is interesting if it is related to concepts which are interesting".

In this architecture, the term **task** refers to specific activities or goals which the discovery system can undertake. Associated with each task is a set of heuristic methods which can be used to execute the task. A typical task is to create a new concept (or conjecture) by specialising (or generalising) an existing concept or conjecture. The methods associated with this task might include the techniques used in concept learning to specialise a classification procedure, and the techniques used in empirical law discovery to specialise laws. Another typical task is to try to find an object which exhibits a particular property. Methods associated with this task might range from general heuristics for designing scientific experiments to formal methods for deducing the object from the definition of the property. Because the methods can be highly specialised, this type of discovery system is called **knowledge intensive.**

The system executes one task at a time, choosing the highest priority task from the list of outstanding tasks, called the **agenda**. The priority of tasks is continuously updated by heuristics similar to those that evaluate the worth of a concept. During the execution of a task heuristics may place new tasks (related to the present task) on the agenda. The concepts and conjectures which are discovered by a system are entirely a function of its tasks, methods and other heuristics. An important aspect of the architecture's open-endedness is that there is no fixed set of tasks or methods.

Two specific systems of this type have been studied. The first, called AM[18], was applied to mathematical domains. The initial set of about one hundred concepts included structures, such as finite sets, lists, and ordered pairs, and operations such as union (of two sets), first element (of a list), equality (of two sets or lists), and composition (of two operations). AM was given roughly two hundred and fifty heuristics in total.

AM quickly discovered elementary numerical concepts such as natural numbers, multiplication, and prime numbers. It also discovered some well-known "conjectures" such as the unique factorisation theorem and Goldbach's conjecture. Some of the concepts discovered by AM were unknown to its author, but were judged by him to be of genuine mathematical interest. An example of this is the concept of maximally factorable numbers.

The successor to AM was Eurisko[17]. The most significant change was to represent heuristics in precisely the same way that concepts were represented. This enabled Eurisko to discover new heuristics using exactly the same mechanism it (and AM) used to discover new concepts.

Eurisko has been moderately successful in several unrelated domains. Its clearest success was in the domain of a wargame involving fleets of hypothetical ships. One fleet is "stronger" than another if it is victorious in battle. Because the battles in this wargame are tactically trivial, the strength of a fleet is entirely a function of its design.

Initially Eurisko was given the concepts and rules of the game and a fleet designed by its author, who had never played the game. It was also given heuristics for modifying ships and fleets, analysing the outcome of simulated battles and so on. Eurisko proceeded to design increasingly stronger fleets. At the same time it was discovering heuristics for designing strong fleets. One very general heuristic that it discovered was "whenever possible, design fleets and ships which are almost, but not quite, extreme in their features". This meant, for example, that fleets should contain almost the maximum number of ships allowed, and that ships that are not heavily armoured should have almost no armour.

With periodic assistance from its author[10] Eurisko designed a fleet that easily won a US national tournament. In a subsequent tournament, changes to the rules meant that the type of fleet that Eurisko had designed was no longer permitted. The general heuristics it had discovered during the design of the first fleet now guided it quickly to a second design satisfying the new rules. This design won the second tournament.

5. References

[1]   Anderson, J. R., J. G. Greeno, P. J. Kline, and D. M. Neves, "Acquisition of Problem Solving Skill," in Cognitive Skills and Their Acquisition, edited by J. R. Anderson, pp. 191-229, Lawrence Erlbaum, 1981.

[2]   Anderson, John R., "Knowledge Compilation: The General Learning Mechanism," in Machine Learning - An Artificial Intelligence Approach, vol. 2, edited by R. S. Michalski, J. G. Carbonell, and T. M. Mitchell, pp. 289-310, Morgan Kaufmann, 1986.

[3]   Angluin, D. and C. H. Smith, "Inductive Inference: Theory and Methods," ACM Computing Surveys, vol. 15, pp. 237-269, 1983.

[4]   Bauer, M. A., "Programming by Examples," Artificial Intelligence, vol. 12, pp. 1-21, 1979.

[10] the author estimates that the final design is 60% his and 40% Eurisko's, and adds that neither one alone could have won the tournament

[5]   Carbonell, J. G., "Derivational Analogy: A Theory of Reconstructive
      Problem Solving and Expertise Acquisition," in Machine Learning – An
      Artificial Intelligence Approach, vol. 2, edited by R. S. Michalski,
      J. G. Carbonell, and T. M. Mitchell, pp. 371-392, Morgan Kaufmann,
      1986.

[6]   Dale, M. B., "On the Comparison of Conceptual Clustering and Numerical
      Taxonomy," IEEE Transactions on Pattern Analysis and Machine
      Intelligence, vol. PAMI-7, no. 2, pp. 241-244, 1985.

[7]   DeJong, Gerald and Raymond Mooney, "Explanation-Based Learning: An
      Alternative View," Machine Learning, vol. 1, no. 2, pp. 145-176, 1986.

[8]   Dietterich, T. G., B. London, K. Clarkson, and G. Dromey, "Chapter
      XIV: Learning and Inductive Inference," in The Handbook of Artificial
      Intelligence, vol. 3, edited by P. R. Cohen and E. A. Feigenbaum,
      William Kaufmann Inc., Los Altos, California, 1982.

[9]   Ellman, Thomas, "Generalizing Logic Circuit Designs by Analyzing
      Proofs of Correctness," in Proceedings of the Ninth International
      Joint Conference on Artificial Intelligence, pp. 643-646, 1985.

[10] Gallant, S. I., "Automatic Generation of Expert Systems From
      Examples," in Proceedings of the Second International Conference on
      Artificial Intelligence Applications, IEEE Computer Society, 1985.

[11] Haase, Ken W. Jr., "Discovery Systems," in Proceedings of the European
      Conference on Artficial Intelligence, pp. 546-555, Brighton, 1986.

[12] Holte, Robert C., "Artificial Intelligence Approaches to Concept
      Learning," in Advanced Digital Information Systems, edited by Igor
      Aleksander, pp. 309-498, Prentice-Hall, 1985.

[13] Huntbach, M. M., "Program Synthesis by Inductive Inference," in
      Proceedings of the Seventh European Conference on Artificial
      Intelligence, pp. 91-100, 1986.

[14] Laird, J. E., P. S. Rosenbloom, and A. Newell, "Chunking in SOAR: The
      Anatomy of a General Learning Mechanism," Machine Learning, vol. 1,
      no. 1, pp. 11-46, 1986.

[15] Langley, P. W., Jan Zytkow, H. A. Simon, and Gary L. Bradshaw, "The
      Search for Regularity: Four Aspects of Scientific Discovery," in
      Machine Learning – An Artificial Intelligence Approach, vol. 2, edited
      by R. S. Michalski, J. G. Carbonell, and T. M. Mitchell, pp. 425-470,
      Morgan Kaufmann Publishers, 1986.

[16] Langley, P. W. and B. Nordhausen, "A Framework for Empirical
      Discovery," in addendum to the Proceedings of the International
      Meeting on Advances in Learning, Rapport de Recherche No. 290, Lab. de
      Recherche en Informatique, Universite de Paris-sud, France, 1986.

[17] Lenat, D. B., "EURISKO: A Program That Learns New Heuristics and
      Domain Concepts. The Nature of Heuristics III: Program Design and
      Results," Artificial Intelligence, vol. 21, pp. 61-98, 1983.

[18] Lenat, D. B. and J. S. Brown, "Why AM and Eurisko Appear to Work," <u>Artificial Intelligence</u>, vol. 23, no. 3, pp. 269-294, 1984.

[19] Michalski, R. S. and R. E. Stepp, "Learning From Observation: Conceptual Clustering," in <u>Machine Learning - An Artificial Intelligence Approach</u>, edited by R. S. Michalski, J. G. Carbonell, and T. M. Mitchell, Tioga Publishing Company, 1983.

[20] Minton, Steven, "Selectively Generalizing Plans for Problem-Solving," in <u>Proceedings of the Ninth International Joint Conference on Artificial Intelligence</u>, pp. 596-599, 1985.

[21] Mitchell, T. M., "Learning and Problem-Solving," in <u>Proceedings of the Eighth International Joint Conference on Artificial Intelligence</u>, pp. 1139-1151, 1983. (Computers and Thought lecture)

[22] Mitchell, T. M., S. Mahadevan, and L. I. Steinberg, "LEAP: A Learning Apprentice for VLSI," in <u>Proceedings of the Ninth International Joint Conference on Artificial Intelligence</u>, pp. 573-580, 1985.

[23] Porter, Bruce W. and E. Ray Bareiss, "PROTOS: An Experiment in Knowledge Acquisition for Heuristic Classification Tasks," in <u>Proceedings of the International Meeting on Advances in Learning</u>, pp. 159-174, Rapport de Recherche No. 290, Lab. de Recherche en Informatique, Universite de Paris-sud, Orsay, France, 1986.

[24] Quinlan, J. R., "Induction of Decision Trees," <u>Machine Learning</u>, vol. 1, no. 1, pp. 81-106, 1986.

[25] Rendell, Larry A., "A New Basis for State-Space Learning Systems and a Successful Implementation," <u>Artificial Intelligence</u>, vol. 20, pp. 369-392, 1983.

[26] Shapiro, Alen D., <u>The Role of Structured Induction in Expert Systems</u>, Department of Artificial Intelligence, Edinburgh University, 1983. (Ph.D. thesis)

[27] Shapiro, E. Y., <u>Algorithmic Program Debugging</u>, MIT Press, 1983.

[28] Silver, B., "Precondition Analysis: Learning Control Information," in <u>Machine Learning - An Artificial Intelligence Approach</u>, vol. 2, edited by R. S. Michalski, J. G. Carbonell, and T. M. Mitchell, pp. 647-670, Morgan-Kaufmann, 1986.

[29] Sleeman, D. H., "Inferring (Mal) Rules from Pupil's Protocols," in <u>Proceedings of the European Conference on Artificial Intelligence</u>, pp. 160-164, 1982.

[30] Smith, Douglas R., "Synthesis of LISP Programs from Examples: A Survey," in <u>Automatic Program Construction Techniques</u>, edited by A. W. Biermann, Gerard Guiho, and Yves Kodratoff, pp. 307-324, MacMillan Publishing Company, 1984.

[31] Vere, S. A., "Inductive Learning of Relational Productions," in <u>Pattern-Directed Inference Systems</u>, edited by D. A. Waterman and F. Hayes-Roth, pp. 281-296, 1978.

10 Learning Diagnostic Rules from Incomplete and Noisy Data

Ivan Bratko and Igor Kononenko

Faculty of Electrical Engineering, E. Kardelj University, Ljubljana, Yugoslavia

INTRODUCTION

It is generally agreed that knowledge acquisition is usually the critical task in the development of an expert system. In the view of this difficulty, machine learning has been for some time considered by developers of expert systems as a promising although still rather remote possibility. However, the methodology of machine learning from examples, based on decision trees, seems to be mature enough for routine applications in many problem areas for which expert systems are typically constructed. This conclusion is supported by a number of other researchers who have been developing this approach and using it in practical applications (e.g. Michie 1986; Quinlan et. al. 1986; Breiman et. al. 1984; Rutowicz and Shepherd 1985) as well as our own results (e.g. Kononenko, Bratko and Roskar 1984).

In this paper we describe some experimental results using the ASSISTANT learning program in medical diagnosis and prognosis, and discuss in particular some features of ASSISTANT which make it robust with respect to errors in the learning data. This sort of robustness is particularly important in application domains in which the data is typically unreliable or noisy, such as medical diagnosis. The effect of noise on the accuracy of learned rules was also studied by Quinlan (1986).

Fig. 1 shows a decision tree for locating the primary tumour in a patient. This tree was generated by ASSISTANT from a set of 170 examples, that is medical records of 170 cancer patients. Internal nodes in the tree are labelled with attributes, that is observable manifestations for this diagnostic problem. The leaves of the tree are labelled with diagnostic classes, that is possible locations of the tumour. Arcs in the tree are labelled with attribute values. For example, the attribute at the root is 'Histological type of carcinoma' which can be either epidermoid (coded 1 in Fig. 1), adeno (coded 2) or anaplastic (coded 3). If the histological type in the patient is either epidermoid or anaplastic then look at the attribute NECK (metastases found in the neck), etc. Notice that the two rightmost leaves of the tree are both labelled 'lung'. This would superficially suggest that the parent node, SEX, of these two leaves is redundant since regardless of its value the diagnosis will be 'lung'. The reason for including SEX at this place in the tree is that the value of this attribute still considerably affects the probability of 'lung', although 'lung' is the most likely outcome in both cases (sex is either male or female). Closer

inspection of this tree printed out in linearised form with
additional information reveals that in the case sex is male 'lung'
is almost certain whereas in the case sex is female lung is the
most likely, but thyroid and vagina are also to be considered. It
should be noted that the tree in Fig. 1 was generated while
applying high 'tree pruning factor'. Pruning will be discussed
later in the paper.

ASSISTANT belongs to the family of tree-based learning programs
that originated from Quinlan's ID3 learning algorithm (Quinlan
1979). Some other members of this family are ACLS (Niblett and
Blake 1981), C4 (Quinlan et.al. 1986), and commercially available
systems ExpertEase, ExTran, RuleMaster (ITL).

It is interesting to notice that outside the AI community, a group
of researchers with statistics background (Breiman et. al. 1984)
has been working on the decision tree approach to machine learning
completely independently from the corresponding effort in AI and
came to similar solutions and conclusions, providing in certain
respects better mathematical evidence.

THE BASIC ALGORITHM

In the type of learning problem we consider in this paper, we have
objects, attributes and classes. Each object belongs to a class.
Objects are described in terms of attributes. An attribute is
either symbolic or numerical. A symbolic attribute has an unordered
set of values. Such a set is typically small, practically never
containing more than ten values. An example of a symbolic attribute
is the sex of a patient. A numerical attribute has an ordered set
of values. An example is the age of a patient, an integer, say,
between 0 and 100. Attributes can be viewed as functions from
objects to attribute values.

The learning task, then, consists of the following:

Given: A set S of objects for learning with specified class
values.

Find: A classification rule that explains the learning set S and
can be used for classifying new objects into classes.

In our case, the decision rule is restricted to the form of a
decision tree.

Objects in the learning set S can be incompletely specified
(unknown attribute values). Also, we will tolerate the possibility
of errors both in attribute values and in class values. Therefore
we say that the learning data can be noisy. We also say that S can
be 'incomplete' in the sense that objects can have unspecified
attribute values. A learning set can be incomplete also in the
sense that it poorly represents the universe of all objects as S
can be very small compared to the complete attribute space. Due to
this kind of incompleteness of S, it may be impossible for the
learning system to generate a reliable rule for classifying new
objects. Noise and incompleteness, of course, make the learning
task more difficult. This sort of difficulty is however typical of

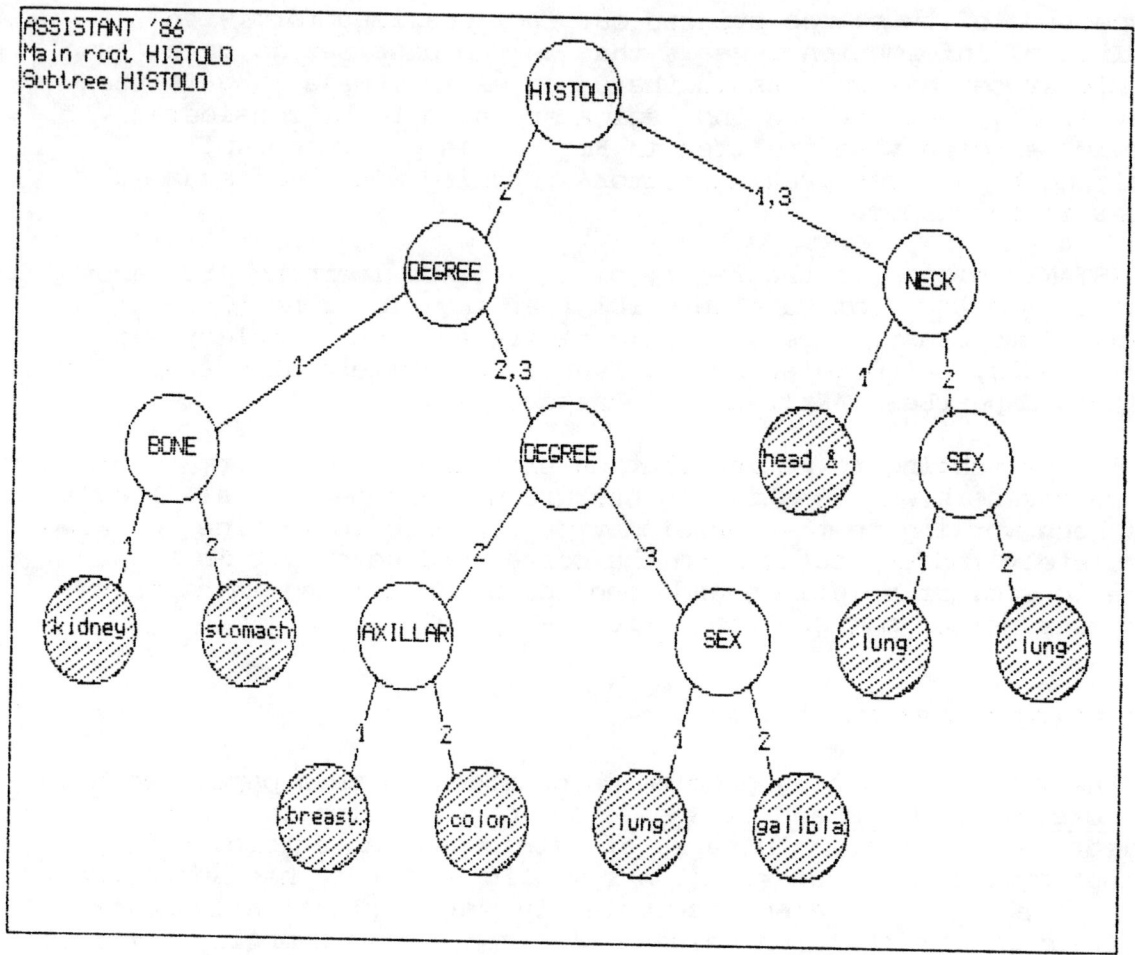

Fig. 1   A decision tree for locating primary tumour in a
cancer patient with metastases as generated and displayed
by ASSISTANT 86, an implementation of ASSISTANT for IBM PC.

Problem domain	non-specialists	specialists	ASSISTANT
Primary tumour	32%	42%	46%
Breast cancer	65%	65%	72%
Lymphography	~ 60%	~ 85%	77%
Thyroid dis.	–	68%	76%

Fig. 2   The accuracy of ASSISTANT's learned rules compared
with the performance of medical doctors - specialists and
non-specialists who are faced with the problem in their
practice. The human's performance for lymphography is a
specialist's estimate whereas the other figures were obtained
experimentally.

many application areas, such as medical diagnosis. The ASSISTANT
learning program that we describe in this paper is specially
equipped with mechanisms for handling noise and incompleteness in
the above sense.

The basic learning algorithm (Quinlan 1979) shared by all the
programs of the ID3 family, and thus also by ASSISTANT, is as
follows (it assumes completely specified examples and symbolic
attributes only):

To construct a rule for a learning set S:

<u>begin</u>
<u>if</u>   all the examples in S belong to the same class, C,
<u>then</u>   the result is a single node tree labelled C
<u>else</u>
    <u>begin</u>
        select the most 'informative' attribute, A,
            whose values are v1, ..., vn;
        partition S into S1, ..., Sn according to value of A;
        construct (recursively) subtrees T1, ..., Tn for
            S1, ..., Sn;
        final result is the tree whose root is A and whose
            subtrees are T1, ..., Tn, and the links between
            A and the subtrees are labelled by v1, ..., vn;
            thus the decision tree generated has the form
            (linearised in the preorder fashion):

            A
                v1: T1
                v2: T2
                ......
                vn: Tn
    <u>end</u>
<u>end</u>

Even in a simplest implementation, this basic scheme needs some
refinements:

(1) We have to specify the criterion for selecting the 'most
informative' attribute. In ID3 it is based on the information
theoretic function. According to this criterion, the most
informative attribute is the one which minimises the 'residual
information', that is the information content left in the example
set after applying an attribute. Residual information of an example
set after applying an attribute A is:

$$\text{Ires}(A) = \sum_V p(v) \sum_C (p(v,c)/p(v)) \log (p(v,c)/p(v))$$

where v stands for values of A, and c stands for classes;
probabilities $p(v)$ and $p(v,c)$ are approximated by statistics on set
S.

(2) If S is empty then the result is a single node tree labelled
NULL (this is the original Quinlan's label, Quinlan 1979).

(3) Each time that a new attribute is selected, only those

attributes are considered that have not yet been used in an upper part of the tree.

(4) If S is not empty and not all the objects in S belong to the same class and there is no attribute left to choose then the result is a single node tree labelled SEARCH (Quinlan's original label again). Label SEARCH means that the set of available attributes is not sufficient to distinguish between class values of some objects (objects that belong to different classes have exactly the same attribute values).

It should be noted that Quinlan (1979) added for efficiency reasons the 'windowing' mechanism to this basic scheme. By that mechanism, if the learning set is very large, a subset called a 'window' is randomly selected and a decision tree generated for this set. This decision tree is tested on the rest of the learning examples. If objects are thus found that are misclassified by this tree then these objects are added to the window and a new decision tree generated for this extended window. These steps are repeated until a perfect decision tree is obtained.

ASSISTANT'S ADDITIONS TO THE BASIC ALGORITHM

Numerous experiments in medical domains (Kononenko 1985) revealed several shortcoming of the basic ID3 algorithm. Here is a summary of the defects:

- There was no provision for handling incompletely specified examples (incomplete information is usual in medical applications).

- ID3 could not properly handle numerical attributes.

- The original ID3's attribute selection criterion seemed to prefer attributes with many values to attributes with fewer values even in cases in which an expert would have little doubt that the fewer-valued attribute was more significant.

-   Decision trees generated tended to be large (often with several hundreds of nodes) and therefore too complex for an expert to understand or to study; thus it was hard to assess whether such a tree captured important regularities of the application domain.

To overcome the drawbacks above, ASSISTANT uses several extra mechanisms added to the basic ID3 algorithm, including those reviewed below:

(1) Handling incompletely specified objects. A learning example with unspecified value for an attribute A is split into a set of examples: so we have for each possible value v of A one example weighed with the probability $p(v|c)$ where c is the class of the example. This probability is approximated by statistics on the learning set. In classification, an object with missing value for attribute A is classified by following in the decision tree all the branches that correspond to all possible values of A, weighed by prior probabilities of the corresponding values of A.

(2) Binarisation of attributes. All the attributes are made

binary. An attribute A is 'binarised' by partitioning its value set into two subsets so that each subset is treated as one value of the resulting binary attribute. Such a subset can be further partitioned into two subsets so that eventually we may end with one-value sets, i.e. with the original values of A. Sets of values of symbolic attributes can be partitioned in any way whereas numerical attributes can only be partitioned into intervals of values so that a subset only contains adjacent values. The point of binarisation is to normalise all the attributes with respect to the number of values. Once the attributes are binarised, it seems, the original information-based criterion for attribute selection gives fair evaluation. Otherwise, this formula gives advantage to attributes with many values although their superiority may only stem from the greater number of values. Therefore such an attribute may only be superior in a 'brute force' way which does not seem fair. The need for normalisation with respect to the number of values has been also observed in (Kononenko, Bratko, Roskar 1984) and Quinlan (1985) where other approaches to normalisation are also discussed and their benefits compared in detail.

(3) <u>Pruning of decision trees</u>. The original ID3 algorithm tends to produce in medical domains very detailed trees with large number of nodes. It can be easily shown that some of this complexity is only the result of noise in the learning data. As an example, consider a situation in which we are to construct a subtree of a decision tree and the current subset of objects for learning is S. Let there be 100 objects in S, 99 of them belonging to class 1 and one of them to class 2. Knowing that there is noise in the learning data and that all these objects all agree in the values of the attributes already selected up to this point in the decision tree, it seems plausible that the class 2 object is in S only as a result of an error in the data. If so, it is best to ignore this object and simply return a leaf of the decision tree labelled with class 1. Since the original ID3 algorithm would in this situation further expand the decision tree, we have by stopping at this point in effect pruned a subtree of the complete ID3 tree. The ASSISTANT program in fact prunes decision trees using a special criterion to decide whether to stop expanding the tree or not. The stopping criterion will be described in the sequel, as well as the effects of pruning.

These additional mechanisms significantly improved the classification accuracy on new objects. In some domains we were able to measure the performance of medical experts themselves. So we could compare the accuracy of ASSISTANT-generated rules with the performance of human experts when they were both working from exactly the same information. Fig. 2 shows results of this comparison. These and some other results suggest a conclusion that ASSISTANT's learned rules can be expected to perform at least comparably to human experts in a typical restricted medical diagnosis problem.

Some of the additional mechanisms of ASSISTANT, pruning in particular, had dramatic effects on the size of generated decision trees, making these decision rules sufficiently small to be easily studied and adopted by human experts. Of course, they can also be easily used for diagnosing without a computer.

TREE PRUNING IN ASSISTANT

ASSISTANT does 'forward' pruning of decision trees. This is similar
to pruning in game playing. A game tree is effectively pruned
during search by simply stopping the expansion of the search tree.
So a 'pruned' subtree is in fact never generated. The critical
question is: When to stop expanding the tree? To decide whether to
stop or not, ASSISTANT in a sense (for analogy with game-playing
programs) performs a one-ply lookahead and estimates whether such a
one-ply expansion of the decision tree would be beneficial or not.
This estimate is done as follows.

Suppose the decision to stop or not is to be made at some current
node t in the tree into which a set S of learning instances falls.
First, the situation is assessed in respect of what would be the
classification accuracy if the tree was terminated at this node. In
this case the resulting leaf would be labelled with the majority
class C in S, and we can estimate the classification error at this
node as $1 - p(C|t)$ where the condition t denotes the event of an
object falling into node t. We will call this error <u>static
classification error</u>. The probability $p(C|t)$ can be estimated by
statistic on S.

The static classification error is to be compared with the error
which would result from applying one more attribute. This error is
estimated using the following algorithm:

        For each example x in S do:
          begin
            S' = S - {x};
            select the best attribute with respect to S';
            extend the tree with this attribute;
            classify x using this extended tree;
            determine the classification error for x
          end
        Sum up the classification errors for all x.

Now compare the estimated average classification error over all x
with the estimated static classification error. If the static error
is lower then stop at node t else continue expanding the tree below
t.

For efficiency reasons the pruning principle above is in ASSISTANT
implemented heuristically. First, the stopping condition is only
evaluated when the number of objects in S is below some threshold.
Second, instead of isolating from S a single element x each time, a
random subset is selected from S for estimating the lookahead
classification error. This is repeated a few times and the error
estimates are averaged.

EXPERIMENTAL RESULTS

ASSISTANT was applied to a number of problems in medical diagnosis
and prognosis. In this section we present some of the results
obtained in the following problem domains:

<u>Prognosis in breast cancer</u>. The classification task here is to

predict whether the disease will recur after the removal of tumour
from breast. Parameters of this learning problem were:

        2 classes
       11 attributes
      286 examples

Diagnosis of lymphatic cancers. The task is to diagnose the results
of the lymphographic investigation. Parameters of this learning
problem were:

        4 classes
       18 attributes
      148 examples

Locating primary tumour. Given a cancer patient with metastases,
the question is what is the location of the original tumour from
which the cancer spread in the body.

       22 classes (i.e. possible locations)
       17 attributes
      339 examples

Urinary tract disorders (female). Given a patient suffering from
incontinence, what is the cause?

        9 classes
       45 attributes
     3661 examples

Thyroid disease. Given the results of clinical findings and
laboratory tests, classify the patient into one of three broad
diagnostic categories.

        3 classes
       39 attributes
      926 examples

In the experiments, for each problem domain the data set was
partioned randomly into a subset of instances for learning
(training set) and a subset for testing. A decision tree was
generated using the training set, and this tree tested for
classification accuracy on the test set. The proportional sizes of
the training set and test set were: 70% of the original example set
was used for training, and the remaining 30% for testing. Breiman
et.al. (1984) comment on the methodological question of what
portion of the data to use for learning and what portion for
testing. They note that usually two thirds of the data are used for
learning and one third for testing although there is no theoretical
support for this particular choice.

Our experimental results on the sizes of decision trees generated
and their accuracy on test data are all the average of 4
experiments with different random partitions of the data into
training and test sets. The classification accuracy is simply the
percentage of correctly classified objects. The size of a decision
tree as given in this paper is the number of leaves in the tree.

Fig. 3 illustrates the effect of binarisation of attributes, comparing the results obtained with the original, multivalued attributes (as in the original ID3) and the results obtained with these attributes binarised. The results in this table were obtained without tree pruning (pruning of decision tree switched off) making it possible to observe the effect of binarisation isolated from other additional mechanisms of ASSISTANT. The empirical conclusion is that binarisation helps both to improve the classification accuracy and reduce the size of decision trees.

Fig. 4 shows the effects of pruning (with binarisation in force). The table compares results with and without pruning. The empirical results clearly confirm the importance of pruning which normally improves the classification accuracy and, often more importantly, greatly reduces the size of the decision tree making the tree more transparent to the user.

CONCLUDING REMARKS

Our practical experience with tree-based learning, also illustrated by examples in this paper, shows that this type of learning methodology can be the basis for powerful practical tools for automated knowledge acquisition for expert systems. The mechanisms added to ASSISTANT make this learning method robust with respect to noise and thus applicable to numerous problem areas in which the data is noisy and incomplete.

The most significant instrument to combat noise is tree pruning. The technique described is <u>forward</u> <u>pruning</u> as it prunes in effect by stopping tree expansion. Although the experimental results are encouraging there is always uncertainty regarding the stopping criterion. Forward pruning is specially sensitive: the stopping criterion based on 1-ply lookahead as in our case may indicate that applying any attribute at some point would be useless; however it may turn out that applying successively <u>two</u> additional attributes would be beneficial. So the pruning may occur too early due to limited lookahead. The problem is the same as in heuristic search where a good solution may be overlooked because heuristically discarding some alternative. In the ASSISTANT system, this uncertainty regarding pruning is aleviated by additional features: the user can adjust certain parameters that affect pruning so that the pruning force can be controlled; also, the user can leash the system at any node and thus also directly overrule the system's decision to prune.

The importance of pruning has been recognised and confirmed in other works in this area (Breiman et.al. 1984; Niblett and Bratko 1986; Quinlan 1986; Rutowicz and Shepherd 1985) where alternative pruning criteria are investigated. More reliable although computationally more complex than forward pruning is 'post pruning' of decision trees. In post pruning, a complete ID3-type decision tree is generated first and then subtrees are pruned in the backward direction according to classification error estimates.

The idea of pruning decision trees has been exported to other type, non-tree based learning. In the AQ15 learning system the pruning of class descriptions is called truncation of covers (Michalski et. al. 1986).

Problem domain	Binarisation on/off	Leaves	Accuracy
Primary tumour	off	140	41%
	on	90	41%
Breast cancer	off	133	67%
	on	63	67%
Lymphography	off	40	75%
	on	22	76%
Urinary tract	off	336	62%
disord. (male)	on	199	66%

Fig. 3   The effect of binarisation of attributes on the size of decision trees and their accuracy. The size of a tree is measured in the number of leaves in the tree.

Problem domain	Pruning on/off	Leaves	Accuracy
Primary tumour	off	90	41%
	on	18	46%
Breast cancer	off	63	67%
	on	9	72%
Lymphography	off	22	76%
	on	14	77%
Urinary tract	off	199	66%
disord. (male)	on	59	67%
Urinary tract	off	357	78%
disord. (female)	on	92	81%

Fig. 4   The effect of tree pruning on tree size and its accuracy.

ACKNOWLEDGEMENT

Our experiments in the medical areas would not be possible without
the assistance of several medical doctors who provided the original
medical data and helped interpreting and cleaning the data. The
data for lymphographic investigation, breast cancer recurrence,
primary tumour and thyroid disease were obtained from the
University Medical Center, Ljubljana; the data for urinary tract
disorders were obtained from Ham Green Hospital, Bristol. We thank
S.Hojker, G.Klanscek, M.Soklic and M.Zwitter of the University
Medical Center, and P.Abrams of Ham Green Hospital. This research
was greatly influenced by exchange of ideas and experience with
D.Michie of The Turing Institute and J.R.Quinlan of The New South
Wales Institute of Technology. We also thank our colleagues
B.Cestnik and E.Roskar for their substantial contributions to this
research.

REFERENCES

Breiman, L., Friedman, J. H., Olshen, R. A, Stone, C. J. (1984)
Classification and Regression Trees. Belmont, California: Wadswordh
Int. Group.

Kononenko, I. (1985) The Design of the ASSISTANT Inductive Learning
System. E.Kardelj University: M.Sc. Thesis.

Kononenko, I., Bratko, I., Roskar, E. (1984) Experiments in
automatic learning of medical diagnostic rules. Technical report:
E.Kardelj Univ., Ljubljana. Presented at ISSEK Workshop 84, Bled,
Yugoslavia 1984.

Michalski, R.S., Mozetic, I., Hong, J., Lavrac, N. (1986) The
multi-purpose incremental learning system AQ15 and its application
to three medical domains. Proc. AAAI 86, Philadelphia, 1986.

Michie, D. (1986) Current developments in expert systems. Proc.
Second Australian Conf. on Applications of Expert Systems, Sydney
1986.

Niblett, T.B., Blake, A. (1981) The ACLS User Manual. Edinburgh
University: Machine Intelligence Research Unit.

Niblett, T.B., Bratko, I. (1986) Learning decision rules in noisy
domains. Expert Systems 86. Cambridge Univ. Press (Proc. Expert
Systems 86 Conf., Brighton 1986).

Quinlan, J.R. (1979) Discovering rules by induction from large
collections of examples. Expert Systems in the Microelectronic Age
(ed. D. Michie) Edinburgh University Press.

Quinlan, J.R. (1985) Decision trees with multi-valued attributes.
Machine Intelligence 11 Workshop. Ross Priory, Scotland, 1985.

Quinlan, J.R. (1986) The effect of noise on concept learning. In

Machine Learning: an Artificial Intelligence Approach, Vol. II
(eds. R.S.Michalski, J.G.Carbonell, T.M.Mitchell). Los Altos,
California: Morgan Kaufmann.

Quinlan, J.R., Compton, P., Horn, K.A., Lazarus, L. (1986)
Inductive knowledge acquisition: a case study. The New South Wales
Institute of Technology, School of Computing Sciences: Technical
report 86.4.

Rutowicz, D., Shepherd, B. (1985) ACLS versus statistical
classifier in a biological application. Machine Intelligence 11
Workshop, Ross Priory, Scotland, 1985.

11 Learning If Then Rules in Noisy Domains

P. Clark and T. Niblett

Turing Institute, Glasgow

1 Introduction

Automatic rule induction systems for inducing classification rules from examples have already proved valuable as tools for assisting in the task of knowledge acquisition for expert systems. In particular, two families of systems based on the ID3 and AQ algorithms have been especially successful. ID3 has been successfully applied to the classification of Chess Endgames (eg. [12], [16], [11]) which were intractable to human experts because of the volume of data, and C4 (an ID3 descendant) to the diagnosis of thyroid diseases [13]. Systems based on the AQ algorithm, such as AQ11 [6] and GEM [15], have been successful in the fields of soya bean diagnosis ([5]) and Chess.

The ID3 algorithm and its descendants share many features in common with Classification trees [1], commonly used by statisticians. The AI community has historically been interested largely in problem domains which are noise-free. Only recently has attention been paid to the problem of learning in noisy domains. The algorithm we present has been designed to produce good rules (in if-then form) when noise is present.

Many of the induction systems mentioned above, developed within the AI community, include methods designed to reduce the effect of noise. One such technique is to perform induction using only *representative* training examples, as selected by the expert or automatically, as was done by the ESEL system [7] for AQ11's application to the task of soya bean diagnosis. A second technique is to use a 'flexible matching' procedure for *applying* the induced rules, whereby their interpretation involves the use of weights and probabilities instead of solely boolean values, thus exploiting to the maximum information contained in the training data. This technique also used by AQ11, and is one of the important features of the AQ15 algorithm [8]. A third method is to alter the rule generation procedure, often involving the relaxing of the constraint that the induced rules should be completely (or to the maximum extent, if completeness is impossible) consistent with the training data. This allows induction to be halted in regions of the search space where there is little training data to guide the system, where further search is as often damaging as beneficial. The pruning of decision trees is an example of this technique, as is done by Assistant (Kononenko et al. 84) and C4 [14]. Quinlan [13] presents a detailed empirical study of the effect of tree pruning in noisy domains. Finally, additional domain knowledge can be used to reduce problems of description language and noise. Explanation-based generalization, (eg. [2], [9]), constrains generalization to be performed only where the generalization's validity can be proved. AQ15 allows the user to provide background knowledge to assist in induction.

This paper presents a description and empirical evaluation of a new induction system based on the third of these techniques, involving the relaxing of the requirement of complete consistency of rules with the training data during their generation. This system, CN2, has been designed with the aim of inducing short, simple, comprehensible rules in domains where problems of poor description language and/or noise may be present. Induced rules are in a form similar to production rules, with the condition being a conjunct of tests on an example's attributes and the conclusion being a class prediction. Thus this paper can be viewed as an investigation of applying a method similar to tree pruning to the generation of 'production rule'-like expressions. These rules are interpreted in a logical manner (not involving weights etc.) in order to maintain their comprehensibility and keep inference well defined for their later use.

In the first part of this paper we describe the CN2 algorithm, and also the other algorithms used for the experimental investigation. In the second part, we present and compare the results of applying CN2 and and a "noiseless" version of the algorithm to two real-world medical domains.

2 Algorithms and Methods

In this section, we first describe the CN2 algorithm and the NL algorithm used for experimental comparison. Secondly we describe the criteria by which we evaluate the induced rules in order to assess the different algorithms.

2.1 Algorithms

We describe the induction systems in terms of five key features. These are

- The form of the induced rules

- The deductive engine for applying the rules

- Operators for moving about the space of rules

- Strategy heuristics for guiding the search of this space

- The domain language used with which to construct rules and describe examples.

In these experiments, the domain languages consisted of attributes, attribute values and classes pre-selected by the user, and were the same for each algorithm. The details of the domain languages used are given in section 3 describing the experiment.

2.1.1 CN2

We now describe CN2, the algorithm proposed in this paper.

Rule Form

CN2 induces rules in a form similar to those generated by the AQ family of induction systems. A *selector* relates an attribute to an attribute value or disjunct of values, eg.

[Weather = wet or stormy]

A selector or conjunction of selectors forms a *complex*. In CN2, a complex defines the condition part of a *decision rule* for identifying a class [1]. A *rule set* is an ordered list of rules for identifying classes (there may be zero or more rules per class). For example, a rule set might be

[Advice = dont_use_umbrella] ⇐ [Weather = sunny]
[Advice = use_umbrella]   ⇐ [Weather = wet ∨ stormy] ∧ [Indoors = no]
[Advice = dont_use_umbrella] ⇐ [Indoors = yes]

Rule Interpretation

To use the rule set to classify new examples, CN2 applies a 'strict match' interpretation by which each rule is tried in order until one is found whose conditions are satisfied by the attributes of the example to classify. The prediction of this rule is then assigned as the class of that example. In this way, the ordering of the rule set is important. The rules could of course be converted to an order-independent form by adding negated selectors to the condition parts.

In the case of no rules being satisfied, a final 'default rule' assigns the class which occurred most frequently in the training examples to the new example to be classified.

Search Technique

During induction, the rule set is gradually built up by generating the rules one by one.

ALGORITHM CN2
1. Search for a satisfactory rule
2. **IF**   a satisfactory rule found
   THEN   add it to the end of the rule set so far
              remove from the training set of examples those which
              satisfy this rule's conditions
              goto 1
   ELSE   stop.

Initially the rule set is empty.

Each rule is generated using the same generation technique, though with a different set of training examples. The training set provided for generating each rule represents those examples *not* satisfying the conditions of any rule in the rule set so far.

The system searches for predictive rules by performing a pruned general-to-specific search. The search is for a complex which is satisfied by a large number of examples of any single class, and few or none of other classes. The best complex found becomes the

---

[1]It is also possible for CN2 to induce a rule set for a *single* class by altering the evaluation function guiding the rule search, eg. by replacing CN2's entropy function (described later) by NL's evaluation function (also described later).

condition part of the new rule, and the most common class of examples it covers becomes the class prediction of the rule.

At each stage in the search, a (size-limited) set or *star* S of 'best complexes found so far' is kept, and specializations of this set examined. Thus the system performs a beam search of the space. The evaluation functions used to bias and prune this search are discussed after the rule generation algorithm is described.

RULE GENERATION ALGORITHM

1. • Form the set S of all *most general* complexes, ie. those containing only one selector.

   • Set 'best rule' to be nil.

2. Evaluate each complex in S. Three types of evaluation are made:

   • What is its quality?

   • Is it statistically significant?

   • Are any specializations of it statistically significant?

   For each complex,

   • If not significant AND cannot become significant by specialization, discard it from S

   • If significant AND better than 'best rule', replace 'best rule' with this complex.

3. **IF** set S is not empty **AND** elements of S can be specialized further

   THEN specialize each complex in S in all ways possible and goto 2

   ELSE stop

If the rule generation algorithm is provided with an empty set of training examples, it immediately fails to find a rule.

Specialization of a complex is performed by either extending it with a new conjunctive term or removing a disjunctive element a selector contains. Each complex can be specialized in several ways, and all specializations are generated and evaluated. Such general-to-specific search has a high branching ratio, and in order to constrain the search the star S of 'best complexes found so far' is limited to a user-defined maximum number of elements *maxstar*. Trimming of the star is performed after completion of step 2, by removing its lowest ranking elements as evaluated by an evaluation function.

One implementation of this specialization process is to repeatedly *intersect* the set of most general complexes with itself, and after each intersection remove all contradictory and unchanged elements in the set.

2.1.2 Heuristics

The above algorithm performs three evaluations, using two evaluation functions. Firstly is the function for assessing rule quality, determining if a new complex should replace the 'best rule' so far, and which complexes in the star S should be discarded if its maximum size is exceeded. To evaluate a complex, the set E of examples which it *identifies* (ie. which satisfy all its selectors) is found and the observed probability distribution $P = (p_1, \ldots, p_n)$

of examples in E amongst classes calculated (where n = number of classes). For example, a given complex may identify 4 examples of class C_1, 2 of C_2, 1 of C_3 and none of C_4, hence $\mathbf{P} = (0.57, 0.29, 0.14, 0)$. Because the search is for complexes identifying a large number of examples of any single class and few of other classes, a function $f = p_{max} - \sum p_{i \neq max}$ (where p_{max} is the largest element of $\mathbf{P}$) would be appropriate. However, it was found that using the information-theoretic measure entropy

$$E = -\sum_{p_i} \log_2(p_i)$$

proved a better function to use (this function to be minimized). The behaviour of E and f are roughly comparable, however entropy will distinguish cases such as $\mathbf{P} = (0.7, 0.1, 0.1, 0.1)$ and $\mathbf{P} = (0.7, 0.3, 0, 0)$ in favour of the latter, a desirable feature as there exist more ways of specializing the latter to a complex identifying only one class (eg. if the examples of the majority class are excluded by specialization, the distributions become $\mathbf{P} = (0, 0.33, 0.33, 0.33)$ and $\mathbf{P} = (0, 1, 0, 0)$ respectively). In addition, entropy tends to direct the search in the direction of more significant rules ; empirically, rules of high entropy tend to also have high significance.

The second evaluation function used is to test whether a complex is *significant*. A complex identifying only examples of one class is an expression of a regularity found in the training data. By 'significant complex' we refer to one which expresses a regularity unlikely to have occurred by chance, ie. reflects a genuine correlation between attribute values and classes in the domain. To assess significance, we compare the *observed* distribution amongst classes of examples satisfying the complex with the *expected* distribution under the null hypothesis that the complex is selecting examples randomly. Some differences in these distributions are to be anticipated owing to random variation. The question we ask is whether the observed distributions are too great to be accounted for purely by chance, ie. that selecting examples at random from the training set would rarely yield the distribution produced by the complex. If so, we consider that the complex is likely to reflect a genuine correlation between attributes and classes.

To test significance, we use the likelihood ratio statistic [4], given by

$$2 \sum_{i=1}^{n} f_i \log(f_i/e_i)$$

where the distribution $\mathbf{F} = (f_1, \ldots, f_n)$ is the observed frequency distribution of examples amongst classes satisfying the complex in question and $\mathbf{E} = (e_1, \ldots, e_n)$ is the expected frequency distribution of the same number of examples under the null hypothesis that the complex is selecting examples randomly. This statistic provides an information-theoretic measure of the (non-commutative) distance between the two distributions (we assume that $\mathbf{F}$ is continuous with respect to $\mathbf{E}$). If $\mathbf{P} = (p_1, \ldots, p_n)$ is the observed *probability* distribution of examples amongst classes satisfying the complex and $\mathbf{Q} = (q_1, \ldots, q_n)$ is the probability distribution of examples in the whole training set, then this measure becomes

$$2N \sum_{i=1}^{n} p_i \log(p_i/q_i)$$

where $N = \sum f_i$ is the total number of examples satisfying the complex, since $e_i = q_i N$. Under suitable assumptions it can be shown that this statistic is distributed approximately as χ^2 with $n - 1$ degrees of freedom. This measure indicates significance –

Class	dog	lion	crab	man
No. of exs. (all)	5	15	10	15
No. of exs. (satisfy complex)	2	13	0	0

Table 1: Distributions of examples amongst classes

the lower it is, the more likely it is that the regularity is due to chance choice of training examples.

In our situation P and Q are found from the observed distributions. We choose a significance threshold α and compute or look up the corresponding significance level for α with the number of degrees of freedom available for classification. Rules are rejected as being insignificant if the likelihood ratio statistic produces a value less than this significance level. In this case we are judging that selecting the same number of examples at random could have led to the observed distribution.

To illustrate the working of this criterion we offer an example, using a significance threshold of 95%. Let there be 4 classes (dog, lion, crab, man) and let the data about distributions be given by table 1.

The significance measure is $2N \sum p_i \log(p_i/q_i) = 2 \times 15 \times (2/15 \times \log 6/5 + 13/15 \times \log 39/15) = 25.57$, and since the 95% significance level for a sample distributed as χ^2 with 3 degrees of freedom is 7.82, we can conclude that this rule is significant (because $25.57 > 7.82$).

Thirdly, a check is made in the algorithm to examine whether specializations of complexes in the star *could* (but not necessarily will be) be significant. It does this by considering the probability distribution P, rather than actually generating such specializations. This check behaves as a fast look-ahead, and is included purely for efficiency. If there are no significant specializations possible, this complex can be discarded.

CN2 is implemented in Quintus Prolog and contains about 300 lines of code, taking approximately 5 minutes run-time to induce a rule set in the lymphography domain (see section 3) on a 4 megabyte SUN-3 using a value of maxstar = 15.

2.1.3   NL (Noiseless)

NL is an induction system using the basic AQ algorithm to generate classification rules. The AQ algorithm is used in a variety of ways by many systems, eg. AQ11 and GEM. Many such systems use this algorithm in a more sophisticated manner than NL to improve predictive accuracy and rule simplicity (for example AQ11 uses a more complex rule interpretation method involving degrees of conformation), hence NL represents a relatively simple AQ-based system.

Rule form and interpretation

NL induces a single decision rule for each class in turn. Each rule consists of a disjunction of complexes (a 'cover') as its condition, and the class as its prediction. The syntax of a complex is the same as described for CN2 earlier, namely a conjunction of selectors.

In NL, a new example is classified by finding which decision rules have their conditions satisfied by the example. If only one rule is satisfied, its predicted class is assigned to the example. If more than one rule is satisfied, then the class to assign is chosen at random from the predictions of those rules. If no rule is satisfied, then the assigned class is by default the one which occurred most frequently in the training examples.

Rule generation

The AQ rule generation algorithm is described well in the literature (eg. [7]), and only a brief overview will be presented here for comparison with CN2. Unlike CN2, NL generates a decision rule for each class in turn. Having chosen a class to generate a rule for, the disjunct of complexes ('cover') forming the condition of the rule is generated in stages, each stage generating a single complex. After generating a complex, the examples covered by it are removed from the training set, and another complex sought for. Complexes are generated until all the examples of the class are covered.

First, an example of the class to generate a rule for is chosen (the 'seed'). Next, an example of a different class (a 'negative example') is chosen, and the most general complexes satisfied by the seed but not by the negative example are generated. This set (or 'partial star') of complexes is then repeatedly specialized to exclude more and more negative examples whilst covering as many positive examples as possible, until all the negative examples are excluded. The best complex is then chosen and returned from the generation algorithm. To make the search tractable, the size of the partial star is limited to a (user-defined) maximum called *maxstar*, the worst elements being discarded should this be exceeded.

Heuristics

The evaluation function used to choose the best complex was "maximize the number of positive examples covered". The function used to trim the partial star size during generation of a complex was "maximize the sum of positive examples covered and negative examples excluded". In the case of a tie for both functions, complexes with fewer selectors were preferred. The seeds were chosen at random. Negative examples were chosen according to their distance from the seed (nearest first, where distance is the number of attributes with different values in the seed and negative example). In the case of contradictions, (when the distance is zero, ie. the seed and negative example have identical attribute values), the negative example is ignored and a different one chosen as the complex cannot be specialized to exclude it but still include the seed.

NL is implemented in Quintus Prolog in about 350 lines of code, taking about 20 minutes run-time to generate a rule set for the lymphography domain on a 4 megabyte SUN-3 using a value of maxstar=15.

2.2   Comparison of CN2 with NL

It is worthwhile to briefly compare CN2 with NL, to indicate some of the main similarities and differences.

CN2 and NL

- CN2 and NL induce rules of similar (but not identical) form, both constructing rules from complexes (ie. conjuncts of selectors).

- The general method of rule generation, namely by repeatedly specializing and evaluating elements of a set ('partial star') of complexes, is used by both CN2 and NL.

This set represents the set of 'best complexes found so far' during the search.

- NL only considers specializing candidate complexes in ways which will cause more negative examples to become excluded from coverage by the complex. NL considers specializing in all ways possible.

- Thus, NL guides its search by comparison of *specific training examples* (the seed and the chosen negative example), and by an evaluation function for pruning the search. CN2 guides its search using an Evaluation function only.

- NL repeatedly specializes complexes until they classify perfectly. CN2 may halt the specialization before this point is reached.

- NL generates a set of rules covering all the training examples. CN2's rules may not cover this entire set.

2.3 Evaluation Criteria

We evaluate the performance of these systems according to two criteria, namely classificational accuracy and rule complexity (to be maximized and minimized respectively). This twofold evaluation is motivated by considering these systems as tools for knowledge acquisition for expert systems. A useful system should induce rules which are both accurate so that they perform well and comprehensible so that they can be validated by an expert and used for explanation.

We do not assess the performance of the algorithms in terms of computer time and memory required to generate and apply their rules, apart from making the observation that they were all able to induce rule sets in the two domains tested in an 'acceptable' time (the run-times for induction were presented earlier with the algorithms, the run-times for classification of 100 new examples using the rules was less than 60 seconds on a 4 megabyte SUN-3 for all the algorithms). We are thus assessing performance on databases typically containing of the order of 500 or less examples, as opposed to much larger machine-generated databases. O'Rorke [11] and [3] provide detailed comparisons of time and memory requirements of ID3 and AQ11P in generating rules from examples using large chess end-game databases.

2.3.1 Classificational Accuracy

Classificational accuracy is assessed by presenting the algorithms with examples not used at induction time for inducing the rules, and measuring the percentage of times that the example's class is correctly predicted by the system (hence the data set is initially split into two parts - training examples and testing examples).

2.3.2 Rule Complexity

For CN2 and NL, we measure complexity by the number of selectors in the final rule set. These measures reveal the gross features of the induced decision rules. More detailed measures of rule complexity have been done by [11] but are not used here.

Property of Domain	Property Value
domain name	lymphography
number of attributes	18
no. of possible values per attribute	between 2 & 8 (average 3.3)
number of classes	4
total number of examples	148
distribution of examples amongst classes	
number of examples in C_1, C_2, C_3, C_4	2,81,61,4
% used for induction (training)	50% (randomly chosen)
% used to test induced rules (testing)	50% (those remaining)

Table 2: Description of the domain lymphography

2.3.3 Combining Accuracy and Complexity

We do not attempt to combine measures of accuracy and complexity, eg. using a function $f = (accuracy - \alpha\ complexity)$, as such a combining function is dependent on the purpose for which the rules are to be used. For example, a 1% fall in predictive accuracy may or may not justify a 10% decrease in rule complexity, depending on the application.

3 Experiment

The two algorithms were tested on two sets of medical data from the domains lymphography and location of primary tumor. The data was obtained from the Institute of Oncology of the University Medical Center in Ljubljana, Yugoslavia. In each test, the training examples were selected at random from the entire data set, and the remaining used for testing. The algorithms were all then run and their rules tested using the same training and testing examples for each algorithm. Ten such tests were performed for each of the three testing conditions (lymphography using 50% of the data for induction, primary tumor using 30%, primary tumor using 70%), and the results averaged.

3.1 The Domains

3.1.1 Lymphography

The properties of this domain are described in Table 2. The data was consistent, ie. examples of any two classes were always different. Both algorithms produced fairly simple and accurate rules, reflecting that the domain was relatively free of noise. This data set was the same as used in [8] to test the AQ15 algorithm.

3.1.2 Location of Primary Tumor

The properties of this domain are described in table 3. The data was inconsistent, ie. examples of different classes existed with identical attribute values. Both algorithms produced more complex and inaccurate rules, reflecting that the domain was relatively noisy. This data set contains fifteen attributes, hence contains slightly less information than that used in [8] which used seventeen attributes.

Property of Domain	Property Value
domain name	location of primary tumor
number of attributes	15
no. of possible values per attribute	between 2 & 3 (average 2.1)
number of classes	15
total number of examples	328
distribution of examples amongst classes	
number of examples in $C_1, \ldots, C_{15}$	82,20,9,14,39,14,6,28,16,7,24,10,29,6,24
% used for induction (training)	30% (test1), 70% (test2)
% used to test induced rules (testing)	70% (test1), 30% (test2)

Table 3: Description of the primary tumor domain

The algorithms were tested using first 30% of the examples as a training set, then 70% as a training set.

4 Results and Discussion

The results of testing were averaged over ten trials. They are shown in table 4. CN2 was tested using two values of significance threshold (as described in section 2.1.1).

From these results, some interesting observations can be made. Most importantly, and perhaps surprisingly, the algorithm designed to reduce problems caused by noisy data achieved a lower rule complexity without damaging predictive accuracy. In the primary tumor domain, for example, using 30% of the examples for induction, CN2 was able to achieve the same classificational accuracy by inducing on average only 8 and 6 short rules with the threshold set at 95% and 99% respectively.

CN2 can be viewed as applying a 'pruning' technique to decision rules expressed in terms of complexes rather than trees. By performing such pruning, 'noisy' algorithms are applying the principle that a sacrifice of classificational accuracy on the *training* data may achieve increased rule simplicity whilst not damaging the predictive accuracy on new test data. This can be seen in Table 5.

AQ15 similarly achieved large reduction in rule complexity without damage to predictive accuracy, compared with complete rule sets generated by the AQ algorithm. However, the methods by which AQ15 and CN2 achieve these results differ. AQ15 first uses an advanced form of the AQ algorithm to generate a rule set. Secondly, a method of *knowledge reduction* is applied which truncates ordered covers and uses them to classify new examples by a technique of *flexible matching*, based on an assessment of the degree of similarity between the example and the condition part of rules. CN2 uses a different rule generation procedure, and may halt specialization of candidate rules in regions of the search space where there are few examples (where further specialization is considered insignificant as described in section 2). These rules are applied to new examples using a 'strict match', examining whether the rule's conditions are satisfied or not by new examples. AQ15's method of cover truncation and CN2's halting of rule specialization can be viewed as applying techniques of 'post-pruning' and 'pre-pruning' respectively, in order to avoid specific poorly-classifying rules being included in the final rule set. As yet, no experimental comparison has been made between these two algorithms.

Domain	Algorithm	Accuracy	Complexity	
Lymphography (50% training)	NL	78%	85	(15 cmpxs, av. 5.7 sels)
	CN2 :			
	(95% threshold)	75%	18	(14 cmpxs, av. 1.3 sels)
	(99% threshold)	75%	14	(10 cmpxs, av. 1.4 sels)
Primary Tumor (30% training)	NL	29%	236	(52 cmpxs, av. 4.6 sels)
	CN2 :			
	(95% threshold)	34%	28	(8 cmpxs, av. 3.5 sels)
	(99% threshold)	32%	19	(6 cmpxs, av. 3.1 sels)
Primary Tumor (70% training)	NL	31%	563	(101 cmpxs, av. 5.6 sels)
	CN2 :			
	(95% threshold)	35%	64	(16 cmpxs, av. 4 sels)
	(99% threshold)	35%	42	(11 cmpxs, av. 3.8 sels)

Table 4: Measurements of rule accuracy and complexity

Domain	Algorithm	Accuracy of rules on training data	Accuracy of rules on testing data
Lymphography (50% training)	NL	100%	79%
	CN2 (95% threshold)	100%	75%
	CN2 (99% threshold)	97%	76%
Primary Tumor (30% training)	NL	83%	29%
	CN2 (95% threshold)	56%	34%
	CN2 (99% threshold)	49%	32%
Primary Tumor (70% training)	NL	69%	31%
	CN2 (95% threshold)	46%	35%
	CN2 (99% threshold)	42%	35%

Table 5: Accuracy of the two algorithms on training and test data

5   Conclusion and Future Issues

This paper has demonstrated that a relatively simple rule induction algorithm is able to achieve a large reduction in rule complexity without damaging classificational accuracy as compared with the other algorithms tested. Especially important is that this result was found in both relatively noiseless and noisy domains. Secondly, the induced rules are in a relatively simple form, similar to the the standard production rule framework used in many expert systems. Both these features are important requirements for practical applications of such an induction system.

Another important result is that the method of significance testing using the likelihood ratio test proved an effective mechanism for controlling search and avoiding regions of the space where examples were sparse.

Currently CN2 requires that an arbitrary confidence parameter is supplied by the user. One direction for future work would be to investigate methods whereby such a parameter can be chosen automatically by the system.

CN2 may halt specialization of rules before they classify the training examples perfectly during rule generation. An alternative approach would be to allow the specialization process to continue until the rules did classify perfectly, and then apply a post-pruning technique to either generalize the rules again or reject them completely. CN2 makes an estimate of the utility of attempting rule specialization (using the significance test as described in section 2). Post-pruning techniques actually perform such specialization and then having done so can evaluate the performance exactly. Post-pruning techniques have been used in AQ15 as mentioned earlier, on ID3 trees [10] and may also be applicable to the CN2 rule generation procedure.

References

[1] L. Breiman, J.H. Friedman, R.A. Olshen, and C.J. Stone. *Classification and Regression Trees.* Wadsworth, Belmont, 1984.

[2] G. DeJong. Generalisations based on explanations. In Kaufmann, editor, *IJCAI-81*, pages 67–69, 1981.

[3] J.A. Jackson. *Economics of Automatic Generation of Rules from Examples in a Chess End-Game.* UIUCDCS-F 85-932, Computer Science Department, Univ. of Illinois at Urbana Champaign, 1985.

[4] J. Kalbfleish. *Probability and Statistical Inference.* Volume 2, Springer-Verlag, New York, 1979.

[5] R. Michalski and R. Chilausky. Learning by being told and learning from examples: an experimental comparison of the two methods of knowledge acquisition in the context of developing an expert system for soybean diagnosis. *Policy Analysis and Information Systems*, 4(2):125–160, 1980.

[6] R. Michalski and J. Larson. *Incremental Generation of VL1 Hypotheses: the underlying Methodology and the Description of Program AQ11.* ISG 83-5, Computer Science Department, Univ. of Illinois at Urbana-Champaign, 1980.

[7] R. Michalski and J. Larson. *Selection of most Representative Training Examples and Incremental Generation of VL1 Hypotheses: the underlying Methodology and the*

Description of Programs ESEL and AQ11. UIUCDCS-R 78-867, Computer Science Department, Univ. of Illinois at Urbana-Champaign, 1978.

[8] R. Michalski, I. Mozetic, J. Hong, and N. Lavrac. The AQ15 inductive learning system: an overview and experiments. In *Proceedings of IMAL 1986*, Université de Paris-Sud, Orsay, 1986.

[9] T.M. Mitchell, R.M. Keller, and S.T. Kedar-Cabelli. Explantion-based generalization: a unifying view. *Machine Learning*, 1(1):47–80, 1986.

[10] T. Niblett and I Bratko. *Learning Decision Rules in Noisy Domains.* TIRM 86-018, The Turing Institute, Glasgow, 1986.

[11] P. O'Rorke. *A Comparative Study of Inductive Learning Systems AQ11P and ID3 using a Chess End-Game Test Problem.* ISG 82-2, Computer Science Department, Univ. of Illinois at Urbana-Champaign, 1982.

[12] J.R. Quinlan. Learning efficient classification procedures and their application to chess end games. In R. Michalski, J. Carbonnel, and T. Mitchell, editors, *Machine Learning: An Artificial Intelligence Approach*, Tioga, Palo Alto, CA, 1983.

[13] J.R. Quinlan. Learning from noisy data. In R. Michalski, J. Carbonnel, and T. Mitchell, editors, *Machine Learning Volume 2*, Tioga, Palo Alto, CA, 1986.

[14] J.R. Quinlan, K.A. Horn, and L. Lazarus. Inductive knowledge acquisition: a case study. In *Proceedings of the Second Australian Conference on the Applications of Expert Systems*, pages 183–204, New South Wales Institute of Technology, Sydney, 1986.

[15] R.E. Reinke. *Knowledge Acquisition and Refinement Tools for the ADVISE Meta-Expert System.* UIUCDCS-F 84-921, Computer Science Department, Univ. of Illinois at Urbana-Champaign, 1984.

[16] A. Shapiro and T. Niblett. Automatic induction of classification rules for a chess endgame. In M.R.B. Clarke, editor, *Advances in Computer Chess*, pages 73–91, Pergammon, Oxford, 1981.

PART D
Statistical Methods in AI

12 Synthesis of AI and Bayesian Methods in Medical Expert Systems

David J. Spiegelhalter
MRC Biostatistics Unit, Cambridge

1 INTRODUCTION

Expert systems research emphasises qualitative reasoning, and powerful structures for representing 'knowledge' have been developed which allow large complex networks of relationships between propositions to be built up, with any 'uncertainty' being treated as an apparently secondary problem often handled in a somewhat ad-hoc manner. With statisticians and engineers becoming more involved in expert systems, the primary role of uncertainty, and hence the need for a soundly-based theory, is being recognised. (See references [1]-[5])

Very briefly, there are five schools of thought: non-numerical methods, fuzzy reasoning, ad-hoc numerical schemes, belief functions, and probabilistic modelling.

1. **Non-numerical methods.**   Perhaps the closest to the aim of simulating human cognition is the argument that qualitative reasoning is adequate for most problems, and more general in its applicability. The work of Cohen [6] on "endorsements" emphasises keeping a trace of evidence for and against propositions of interest, without numerical weighting of their relative importance, but with explicit recognition of the quality of underlying evidence on which the association is based. It is claimed this is missing from a quantitative analysis.

2. **Fuzzy reasoning.**   Zadeh [7] claims that much of what goes under the label of "uncertainty" may be expressed as numerical representations of imprecise linguistic terms, such as "likely", "very likely" and so on. Large medical systems, acknowledging the vague nature of many medical terms, have been constructed on this principle. If f(a) represents the possibility of a proposition a, the usual rule for possibilities of the conjunction or disjunction of propositions is that f(a and b) = min(f(a),f(b)); f(a or b) = max (f(a),f(b)).

3. **Ad-hoc numerical methods.**   Early systems such as MYCIN,
    PROSPECTOR, CASNET, INTERNIST, adopted quasi-probabilistic but
    essentially ad-hoc methods of weighing and combining evidence, and
    these have been perpetuated in the expert system shells that have
    been developed. This has resulted in little practical development
    in this particular facet of expert systems over a decade, reflecting
    the low priority the majority of workers in artificial intelligence
    place upon formal evaluation criteria for the internal processes of
    a system or its prediction of future events. In fact, normative or
    prescriptive criteria often appear to be deliberately criticised as
    irrelevant in AI research. Hajek [8], however, has established a
    formal mathematical framework for any numerical calculus for
    combining and propagating evidence, which shows the essential
    isomorphism of such calculi under fairly broad conditions.

4. **Belief functions.**   The problems in assigning point
    probabilities to all propositions in a system has led people to
    investigate interval valued beliefs, and belief functions [9]appear
    to have many intuitively desirable properties. By placing
    probability mass over the set of all subsets of propositions, they
    enable one to 'reserve judgement' over the precise recipient of
    one's belief, while the Dempster rule of combination is intended to
    allow disparate sources of evidence to be amalgamated. However, the
    cost is great computational complexity and an apparent lack of
    criteria for external calibration of the expressions of belief in
    order to allow assessment, learning, criticism or decision-making.

5. **Probabilistic modelling**   Although accused of being
    'epistemologically inadequate' for the challenge posed by expert
    systems, probability has generally been rejected largely on the
    purely pragmatic grounds of being too complex to implement
    correctly, and the unavailability of appropriate quantities. This
    view has been challenged by Cheeseman, Pearl, and others [2], who
    have pointed out a variety of techniques for limiting the number of
    assessments that have to be made and providing rigorous, but
    efficient, algorithms for updating beliefs in the light of incoming
    evidence.

In the following sections, we sketch out a strategy for dealing with
knowledge bases of arbirtrary size, but whose graph is fairly sparse.
Pearl [2] discusses the appropriateness of this assumption in many
fields.

2   KNOWLEDGE REPRESENTATION AS A CAUSAL NETWORK

A directed graph is a common structure for representing, in a broad
sense, 'causation'.   Often there is a natural ordering to unknown
characteristics of an individual patient, in which A precedes B, say, if
an expert would consider it crucial to know the value that A took on
before assessing the chance of a particular value of B occurring.   Thus
a causal network may be built up, in which a node of the graph may have
a number of direct influences or 'parents'.   Translating such
<u>qualitative</u> assessments into a probabilistic model requires independence
assumptions related to path analysis [10]; essentially, if we know the
values of the <u>parents</u> of a node B, then our opinion about B is unchanged
by any other evidence not concerning descendants of B.   With a minimal
number of <u>quantitative</u> probability assessments, this then allows a full
joint distribution over the entire network to be represented in a

recursive form corresponding to the initial ordering on the graph [10].
When evidence is received concerning a node in the graph, the effects
need to be propagated through the network, which will generally entail
going against the original directed probability assessments.   Kim and
Pearl [11] developed an elegant algorithm when the graph forms a tree
structure - that is, when removal of any edge of the graph breaks it
into two. However, keeping within the directed representation seems to
make propagating through general networks somewhat clumsy [12,13]. An
undirected representation of our belief structure appears necessary, and
this is discussed in the next section.

3   THE USE OF UNDIRECTED REPRESENTATIONS

Fortunately there exists a theory of undirected graphs representing
qualitative belief structures, based on Markov random field models [14]
[15].   Relevant details are provided in references [10], [16], which
give the link between directed and undirected models.

As described in [17], the topology of the directed graph needs to be
changed by adding certain dummy constraints and dropping the directions
on the edges: the more complex undirected graph remaining demonstrates
certain independence assumptions based on a separability criteria [10],
which allow a simple representation of our beliefs. Certain of the
original independence assessments are no longer explicitly represented
in the stored graph, but are still implicit in the numerical assessments
made.

We shall illustrate the adaptations using Figure 1 as an example, in
which two independent risk factors a,b affect an underlying disease c,
which may cause a symptom f through two physiological pathways, d,e.

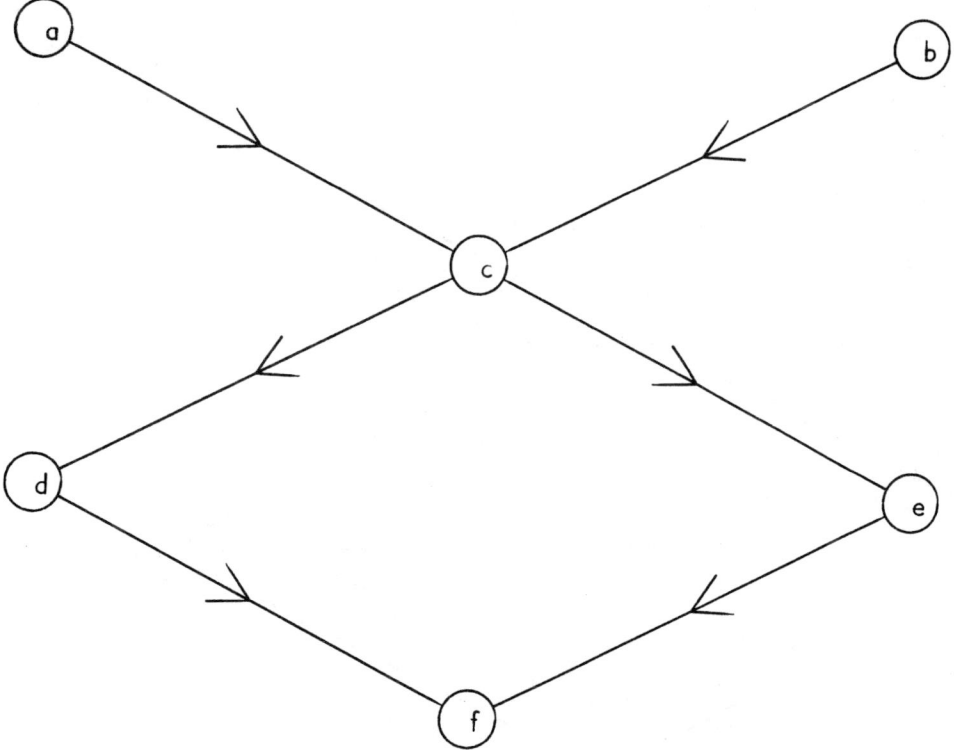

Figure 1: Original directed graph : a;b; etc. are binary variables
          taking on values a_0, a_1; b_0, b_1; etc.

To specify our beliefs on the causal networks in Figure 1, we need to specify probabilities $p(a_1)$, $p(b_1)$, $p(c_1|a,b)$, $p(d_1|c)$, $p(e_1|c)$, $p(f_1|d,e)$, from which probabilities for a_0, b_0 etc. are obtained by subtraction. (The term '$p(c|a,b)$', for example, is used as shorthand for $p(c_1|a_0,b_0)$, $p(c_1|a_0,b_1)$, $p(c_1|a_1,b_0)$, $p(c_1|a_1,b_1)$, <u>all</u> of which need to be assessed to give the full conditional distribution). Our joint belief in a particular combination of circumstances $a*,b*,c*,d*,c*,f*$ is then simply

$$p(a*,b*,c*,d*,e*,f*)$$

$$= p(a*)\ p(b*)\ p(c*|a*,b*)\ p(d*|c*)\ p(e*|c*)\ p(f*|d*,e*) \qquad (1)$$

The problem with the representation (1) is in calculating particular marginal and conditional probabilities. Suppose, for example, we observe $f* = f_1$ (i.e. the symptom is present) and we want to work out our new belief that risk factor a is present, $p(a_1|f*)$; a long, cumbersome and opaque series of manipulations on (1) are required [13]. However, Kiiveri, Carlin & Speed [10] and Wermuth and Lauritzen [16] show that if such a recursive representation has no 'unjoined parents' (in the sense that a,b are unjoined parents of c in Figure 1), then the beliefs may also be represented in undirected form in terms of marginal distribution on the 'cliques' (maximum subgraphs in which all nodes are joined) of the undirected graph formed by dropping the arrows. Furthermore this graph is 'triangulated', in the sense of having no cycles of length 4 or more without a 'short-cut'. Lauritzen and Spiegelhalter [19] go on to describe algorithms for 'marrying' parents of a directed graph such as Figure 1 (i.e. adding edges a-b and d-e), then 'filling in' to triangulate using the work of Tarjan and Yannakakis [18].

In Figure 2 we show the 'moral graph' formed by marrying parents in Figure 1, which does not require a fill-in as it is already triangulated.

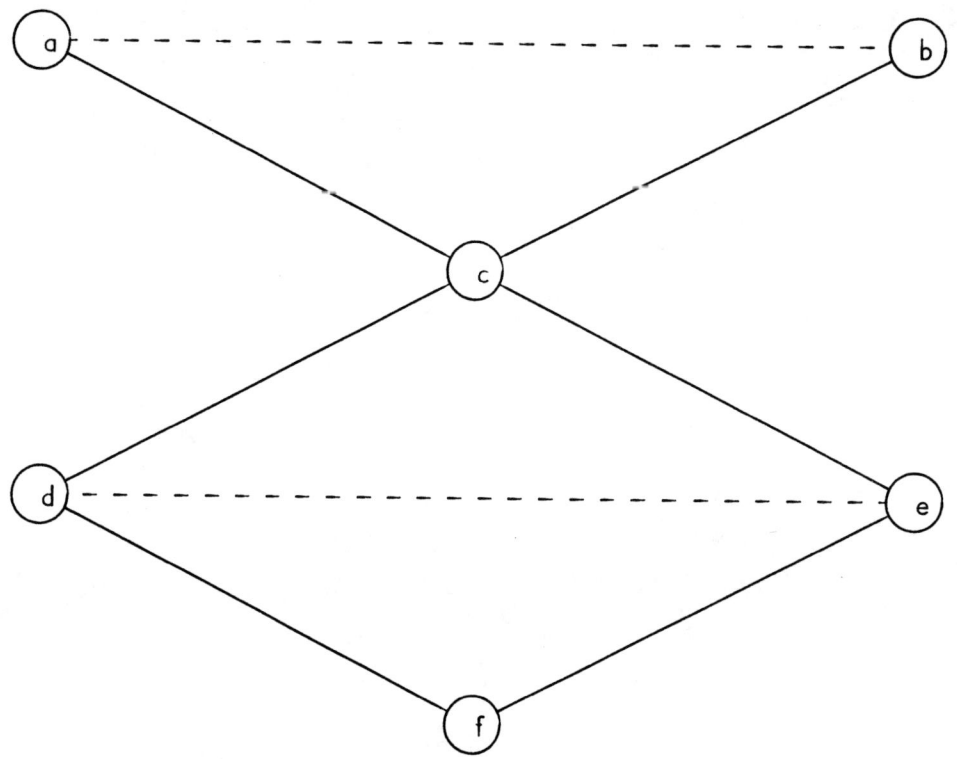

Figure 2.   Moral graph formed from Figure 1, with cliques [a,b,c], [c,d,e], [d,e,f].

A model such as in Figure 2 is called 'decomposable' and we shall show
how our beliefs are decomposed into marginals on cliques and clique
intersections. From our original assessments, we can calculate $p(a,b) =$
$p(a) p(b)$, and $p(d,e|c) = p(d|c) p(e|c)$. This allows (1) to be written

$$p(a*,b*) \ p(c*|a*b*) \ p(d*,e*|c*) \ p(f*|d*,e*)$$

$$= \frac{p(a*,b*,c*) \ p(c*,d*,e*) \ p(f*,d*,e*)}{p(c*) \qquad p(d*,e*)} \ . \qquad\qquad (2)$$

We have thus obtained an undirected representation of our beliefs for
use by the machine, where the user has had to make no additional inputs,
and can still be confronted with his original directed model on the
screen. This internal, undirected model allows efficient algorithms for
evidence propagation, as we shall see in the next section.

4.   EVIDENCE PROPAGATION

Suppose we now observe that $f* = f_1$, and we wish to propagate the effects
of the evidence through the entire network in a single pass using local
computations. We have previously described [17] how algorithms [18] may
be used to order the cliques of the graph starting with the one containing
the observed node. Specifically, 'maximum cardinality search' [18]
numbers the observed node '1', then iteratively numbers as the next node
that adjacent to the maximum number of previously numbered nodes (breaking
ties at random). Cliques are then ordered according to their 'maximum'
node. Figure 3 shows such an ordering.

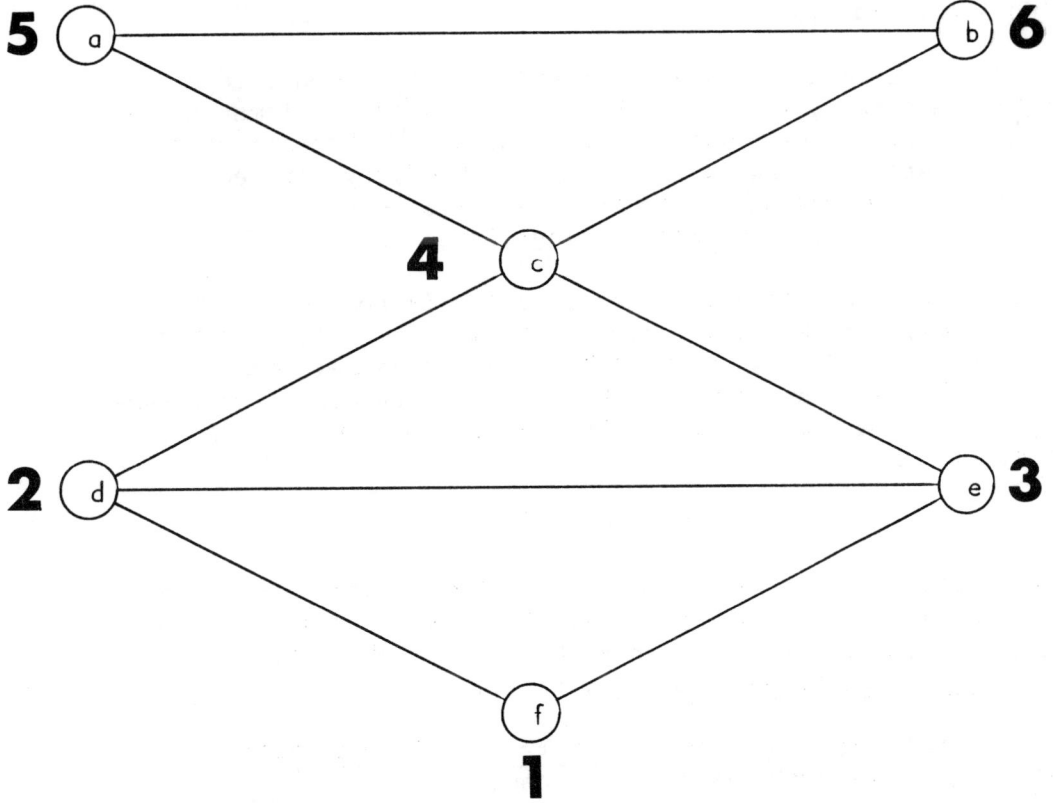

Figure 3:   Ordering of nodes by 'maximum cardinality search' [18]
             following observation of node f.

Let us denote updated probabilities by p^*.   Then for the first clique $[d,e,f]$ we have $p^*(d,e) = p(d,e|f_1)$ which is trivially obtained from the marginal $p(d,e,f)$.   For the next clique $[c,d,e]$ we required

$$p^*(c,d,e) = p(c,d,e|f_1)$$

$$= p(c|d,e) \, p(d,e|f_1) \qquad \text{since } d,e \text{ separate } c \text{ from } f$$

$$= p(c,d,e) \, p^*(d,e) \, / \, p(d,e)$$

This is an example of the general updating formula [17], that states that our revised belief on a clique is given by our previous belief times the ratio of revised to previous beliefs on the clique intersection.   Having updated our beliefs on $[c,d,e]$, we then propagate to the final clique $[a,b,c]$ by

$$p^*(a,b,c) = p(a,b,c) \, p^*(c) \, / \, p(c),$$

and hence obtain $p^*(a)$.

The observed node can then be dropped from the graph leaving a new decomposable model ready to receive and propagate evidence concerning any node.

5   CONCLUSIONS

The strategy described here may be implemented in the more expensive expert system shells.   There are, however, a number of issues still to be adequately tackled.

The first concerns 'logical links', in which certain conditional probabilities are zero, which in theory could cause problems in the algorithms, but with care should be surmountable.   Perhaps most important is a straightforward mechanism for handling 'imprecise probabilities' which would seem to require a simple Bayesian methodology for a particular class of log-linear model.

Other problems include reasonable search criteria for important questions to ask, and adequate explanation facilities for revised beliefs.   In spite of these practical tasks, we believe that a coherent probabilistic structure is the appropriate model for future expert systems research, and continuing developments [19] suggest feasible solutions.

REFERENCES

1   Spiegelhalter D J and Knill-Jones R P (1984) Statistical and knowledge-based approaches to clinical decision-support systems, with an application in gastroenterology (with discussion). J. Royal Statistical Soc., B, 147, 35-77.

2   Kanal L N and Lemmer J (1986) Uncertainty in Artificial Intelligence. North-Holland, Amsterdam.

3   Cheeseman P (1985) In defense of probability.   IJCAI-85, 1002-1009.

4   Henrion M (1987) Should we use probability in uncertain inference systems? Proceedings of 8th Annual Conference of the Cognitive Science Society, Lawrence Erlbaum: Amherst.

5   Spiegelhalter D J (1986b) Probabilistic reasoning in predictive
    expert systems. In Uncertainty in Artificial Intelligence, (eds
    Kanal, L N & Lemmer, J) North-Holland: Amsterdam. pp 47-68.

6   Cohen P R (1985) Heuristic Reasoning about Uncertainty: an
    Artificial Intelligence Approach, Pitmans: Boston.

7   Zadeh L A (1983) The rôle of fuzzy logic in the management of
    uncertainty in expert systems. Fuzzy Sets and Systems, 11,
    199-228.

8   Hajek P (1985) Combining functions for certainty degrees in
    consulting systems. Technical Report, Mathematical Institute,
    Prague.

9   Shafer G (1976) A Mathematical theory of evidence. Princeton
    University Press.

10 Kiiveri H, Speed T P, Carlin J B (1984) Recursive causal models.
    J. Austral. Math. Soc. (Series A), 36, 30-52.

11 Kim, J H and Pearl, J (1983) A computational model for causal
    and diagnostic reasoning in inference systems. Proceedings 8th
    International Joint Conference on Artificial Intelligence,
    Karlsruhe, West Germany : p190-193.

12 Pearl J (1986) Fusion, propagation and structuring in belief
    networks. Artificial Intelligence, 28, 9-15.

13 Shachter, R D (1986) Intelligent probabilistic inference.
    In Uncertainty in Artificial Intelligence, (eds Kanal, L N &
    Lemmer, J) North-Holland: Amsterdam. pp 371-382.

14 Isham V (1981) An introduction to spatial point processes and
    Markov random fields, International Statistical Review, 49, 21-43.

15 Darroch J N, Lauritzen S L and Speed, T P (1980) Markov fields
    and log-linear models for contingency tables.
    Annals of Statistics, 8, 522-539.

16 Wermuth N and Lauritzen S L (1983) Graphical and recursive
    models for contingency tables. Biometrika, 70, 537-52.

17 Spiegelhalter D J (1987) Coherent evidence propagation in expert
    systems. Statistician, (to appear).

18 Tarjan R E and Yannakakis M (1984) Simple linear-time algorithms
    to test chordality of graphs, test acyclicity of hypergraphs, and
    selectively reduce acyclic hypergraphs.
    SIAM J. Comput. 13, 566-579.

19 Lauritzen S L and Spiegelhalter D J (1987) Fast manipulation of
    probabilities with local representations - with applications to
    expert systems. Technical Report 87-7. Institute of Electronic
    Systems, Aalborg University.

13 An Introduction to Statistical Pattern Recognition

B. D. Ripley

University of Strathclyde, Glasgow

1   WHAT IS PATTERN RECOGNITION?

Like most scientific disciplines pattern recognition (hereafter PR) is not well-defined and the term is in fact used in more than one sense in different communities. The "patterns" to be recognized can be quite general. Some specific examples will help:

>   recognizing handwritten numerals
>   comparing fingerprints with a database of images
>   identifying an attacking missile from radar signals
>   classifying landuse in a satellite image
>   prognosis of head-injury cases from medical tests
>   recognizing chemical compounds from gaschromatography

All these problems involve making a decision on the basis of a structured multidimensional signal. The decision is to classify the signal as coming from one of a finite number of sources, or perhaps to report "don't know" or "not a class known to me". The problem is to design a classifier which when presented with a signal will make a decision reliably and (often) fast. To design a classifier one is usually given a few samples of signals from each class, the training set. Frequently there is a second training phase in which the prototype classifier is corrected, and this can be used to tune the classifier automatically. This could be termed a learning phase. Finally the classifier is used without further supervision.

Pattern recognition as described above is sometimes called supervised learning. This is in contrast to unsupervised learning, which means cluster analysis. The electrical engineering community distinguishes two (or three) types of PR. So-called statistical pattern recognition is concerned with the actual decision-making process, and is virtually coincident with the statistical subject of discrimination and classification.   Structural pattern recognition is concerned with describing the classes to be discriminated in a compact way. For example, we could try to describe handwritten numerals by the strokes used, and there are conventional ways to describe fingerprints. Within structural pattern recognition, syntactic PR describes such structures by formal grammars, so encapsulating the structural description in formal rules.

Although the aims of statistical PR and parts of multivariate analysis are very similar, the emphases differ somewhat. Until recently statisticians have been concerned with small sets of data and a few measurements on each object. Fisher's overused Iris data [5] provide an example: a training set of fifty objects (plants) in each of three classes (species) with four measurements on each object. Many biological examples are smaller. In contrast, in PR applications there may be hundreds of measurements on each object, tens of classes and a few objects of each class in the training set. An example is training a classifier to read handwritten symbols on maps. Part of the design problem may be to select just a few measurements to be made on future objects. Because pattern recognition procedures are to be used without human supervision, it is important that they are robust to departures from assumptions, and able to report weak decisions where appropriate. Sometimes it may be appropriate to refuse to decide. In these respects PR systems must behave like an expert human classifier and could be said to display artificial intelligence. In its early days artificial intelligence meant automated pattern recognition (e.g. Minsky, [12]).

2   BASIC METHODS OF STATISTICAL PR

Statistical PR, almost by definition, has probability models. Suppose objects can be of class $\omega_1,\ldots,\omega_c$, and on each object we observe a vector **x** of observations, known as the feature vector. We assume that the population of members of class ω_i gives rise to feature vectors with pdf $p(\mathbf{x}|\omega_i)$. Note that the randomness here can arise from two sources; it reflects both measurement errors and the differences between members of the class.

2.1 The Bayes rule

To formulate the Bayes rule we need an additional component, the prior probabilities $\pi_1,\ldots,\pi_c$ of the object belonging to class $\omega_1,\ldots,\omega_c$. Then the Bayes rule is to choose that class which maximizes the posterior probabilities $\pi(\omega_i|\mathbf{x})$. These probabilities are of course computed as

$$\pi(\omega_i|\mathbf{x}) = \frac{\pi(\omega_i)p(\mathbf{x}|\omega_i)}{\sum_j \pi(\omega_j)p(\mathbf{x}|\omega_j)}$$

by Bayes' theorem, so we choose ω_i to maximize $\pi(\omega_i)p(\mathbf{x}|\omega_i)$. In the language of decision theory, the Bayes rule here is the Bayes rule for the loss function

$$L(\text{choose } \omega_i,\ \omega_j \text{ true}) = \begin{cases} 1 & i \neq j \\ 0 & i = j \end{cases}$$

so all incorrect decisions are penalized equally.

It is generally accepted that the Bayes rule is best if all the quantities required are known and it is feasible (i.e. can be computed in the time available). In practice $p(\mathbf{x}|\omega_i)$ will be estimated from the training set, and $\{\pi_i\}$ from past data. In particular, if $p(\mathbf{x}|\omega_i)$ is estimated parametrically (say by a multivariate normal distribution), small errors in the fit to the parametric family can make large changes to the decision rule. Consider just two classes ω_1 and ω_2, with scalar **x** and

$$p(\mathbf{x}|\omega_1) = f(x), \qquad f \text{ unimodal about zero}$$

$$p(\mathbf{x}|\omega_2) = f(x-\mu), \qquad \mu > 0$$

Then the decision rule is

$$\text{choose } \omega_2 \text{ if } \frac{f(x-\mu)}{f(x)} > \frac{\pi_1}{\pi_2}$$

and typically the decision boundary will be at a value of x where f(x) and f(x-μ) are both small. Thus assumptions about the tail behaviour of $p(\mathbf{x}|\omega_i)$ will govern the decision boundary. In multivariate problems we will have essentially no information on this tail behaviour!

2.2 The kNN rule

At first sight this is a sensible <u>ad hoc</u> suggestion. In some suitable distance metric, pick the k nearest points in the training set to the feature vector **x** of the object to be classified. If there is a majority of points from one class, decide on that class. If not, choose amongst the most represented classes on the basis of the distance to the points. For example, one might choose the class with the smallest averaged squared distance.

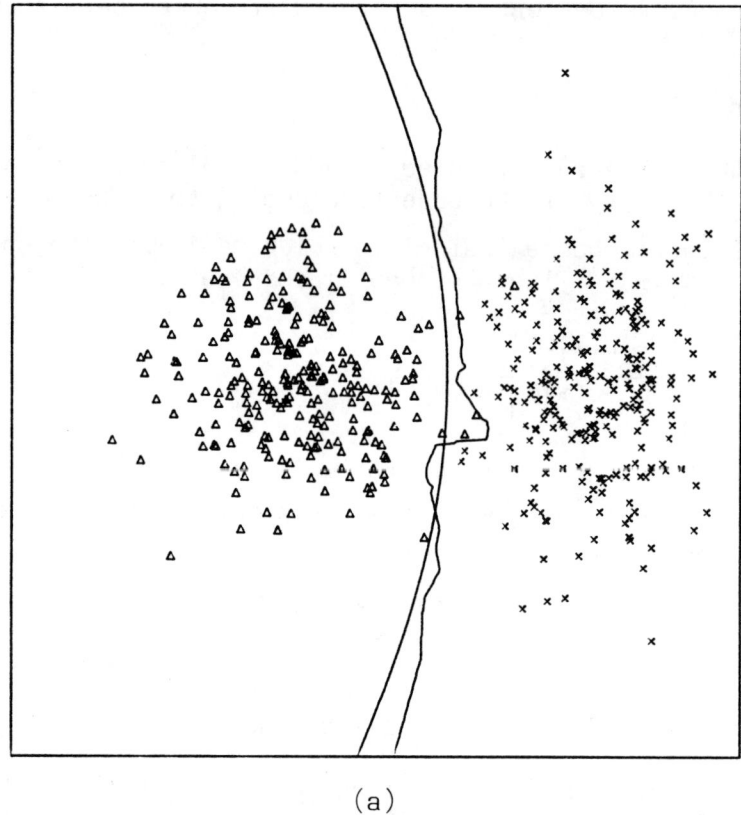

(a)

FIG. 1   Two samples of 250 points each from normal distributions with different covariances. The hyperbolic curve is the Bayes rule decision boundary. (a) 3NN rule based on the whole sample. (b) 3NN rule based on about one quarter of the samples and (c) based on 10% of the samples. The numbers are n_1 and n_2 of the edited training sets.

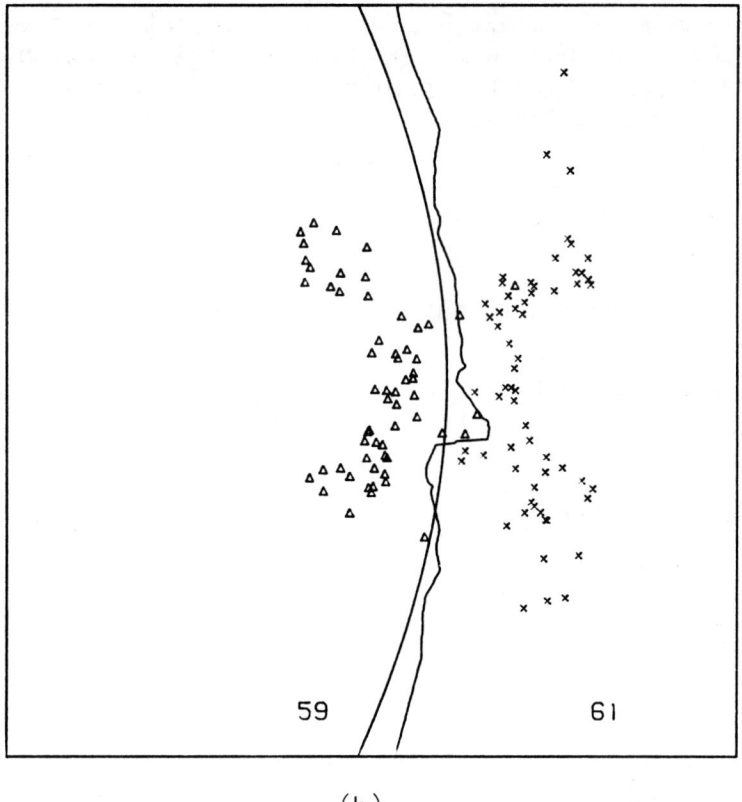

(b)

(c)

This method has the advantage of potential speed. As figure 1 shows, we may not even need all the training set, for an edited subset may give the same (or nearly the same) rule. Further, an asymptotic analysis of its performance is possible under weak conditions. This is asymptotic in the size n of the training set, with training samples drawn independently from the whole population of objects. The average error rate of the Bayes rule is

$$E^* = \int \{1 - \max_i \pi(\omega_i | \mathbf{x})\} p(\mathbf{x}) d\mathbf{x}$$

where $p(\mathbf{x}) = \sum_j p(\mathbf{x}|\omega_j)\pi_j$ is the pdf of $\mathbf{x}$ in the whole population.   Let E_k denote the average error rate of the kNN rule, that is the probability that it makes an error. It can be shown that, asymptotically,

$$E^* \leq E_1 \leq E^*(2 - \frac{c}{c-1} E^*)$$

and

$$\lim_{k \to \infty} E_k = E^*$$

These asymptotic results can be misleading about finite training sets. They are not surprising when the kNN rule is viewed as an estimated Bayes rule. Suppose we decide to use the k samples nearest to $\mathbf{x}$ to estimate the $p(\mathbf{x}|\omega_i)$.   Natural estimates are

$$\hat{p}(\mathbf{x}|\omega_i) = \ell_i(\mathbf{x})/n_i v(\mathbf{x})$$

where $v(\mathbf{x})$ is the volume of the ball containing the nearest k points, and $\ell_i(\mathbf{x})$ of those points correspond to class ω_i.   Since n_i/n is a natural estimate of π_i, we conclude

$$\hat{\pi}(\omega_i|\mathbf{x}) = \ell_i(\mathbf{x})/k$$

and with this estimate the Bayes rule <u>is</u> the kNN rule.

This may seem an unorthodox way to estimate a density, but it has been used (Tukey and Tukey, [19]; Silverman, [18], §§2.6 and 5.2).   If computing time is no problem then better methods of non-parametric density estimation are available, and these should give better approximation to the Bayes rule (§2.4).

The kNN rule is widely used for its simplicity. Its asymptotic properties have perhaps lulled its users into a false sense of security, and large training sets, carefully edited, do seem desirable. It has the attraction of a fully automatic method; once k is chosen the editing and hence the classifier design can be totally automatic.

If a firm decision is not essential, the kNN rule can be modified to decide on class ω_i only if at least ℓ of the k nearest points belong to class ω_i.   If this is satisfied for no class ω_i, a doubt decision is recorded.   Again an asymptotic analysis is possible (Devijver and Kittler,

[4], Chapter 3) and the modification seems to be quite robust.

The kNN rule was originally proposed in statistics, by Fix and Hodges [6].

2.3 Linear separability

Suppose we have c classes and multivariate normal distributions $\mathbf{x} \sim N(\boldsymbol{\mu}_i, \Sigma)$ for class ω_i, i = 1,...,c. Then the Bayes rule amounts to choosing the class with the smallest value of the Mahalanobis distance

$$\Delta_i = (\mathbf{x}-\boldsymbol{\mu}_i)^T \Sigma^{-1} (\mathbf{x}-\boldsymbol{\mu}_i)$$

If c = 2 then

$$\Delta_2 - \Delta_1 = -2(\boldsymbol{\mu}_2-\boldsymbol{\mu}_1)^T \Sigma^{-1} [\mathbf{x}-\tfrac{1}{2}(\boldsymbol{\mu}_1+\boldsymbol{\mu}_2)]$$

so the decision boundary is a hyperplane perpendicular to the line joining $\boldsymbol{\mu}_1$ and $\boldsymbol{\mu}_2$. We have of course rederived Fisher's linear discriminant analysis, [5].

Early work in pattern recognition looked for linear separability for its own sake. That is, a linear discriminator $\mathbf{a}^T\mathbf{x}$ was sought so that $\mathbf{a}^T\mathbf{x} < C$ for all members of class ω, and $\mathbf{a}^T\mathbf{x} > C$ for all members of class ω_2. This is only going to be possible if the $p(\ |\omega_i)$ have bounded support, and $\{\mathbf{x}|p(\mathbf{x}|\omega_1) > 0\} \cap \{\mathbf{x}|p(\mathbf{x}|\omega_2) > 0\} = \emptyset$. Even then <u>linear</u> separation may not be possible. It is more plausible that the training set could be linearly separated, and this was the aim of early methods such as Nilsson's linear learning machine [14]. Suppose we have n points in the training set. These can be divided in 2^n ways between the two classes, some linearly separable. Let d denote the dimension of the feature space (set of feature vectors). Then if n < 2(d+1) most divisions are separable, whereas for n > 2(d+1) few are. This critical phenomenon is quite sharp for moderate to large d, and suggests that we may expect to fail to linearly separate with n > 2(d+1) <u>unless</u> the classes ω_1 and ω_2 are well separated. Given that a separating linear hyperplane exists, Nilsson's LLM attempts to find it by successive approximation. A similar algorithm called the Perceptron became famous in the 1960's (Rosenblatt, [17]; Minsky & Papert, [13]).

Thus far we have concentrated on separating two classes. A number of extensions to c > 2 classes are possible. The most elegant is to set up a linear discriminant function $g_i(\mathbf{x}) = \mathbf{a}_i^T\mathbf{x} + b_i$ for each class, and choose the class for which $g_i(\mathbf{x})$ is maximal. As figure 2 illustrates, this divides the feature space into a polytope corresponding to each class.

In some problems it will be obvious that linear separability is inadequate. One can try transforming the feature space to suit a linear classifier. Such techniques are considered further in section 3.

It seems that linear separability methods are now little used. They were developed for computational speed, and in some romantic hope of modelling the behaviour of the human brain as it was perceived 25 years ago. Their terminology still influences the jargon of pattern recognition.

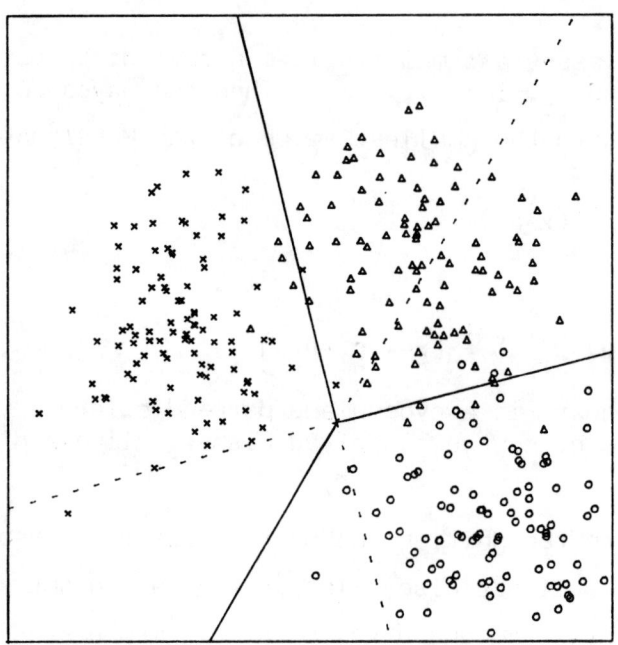

FIG.2 Linear discriminant boundaries for three classes divide the feature space into polytopes. The dashed lines are unused parts of the decision boundaries between two classes.

2.4 Kernel discriminant analysis

The kNN rules learn about the class-conditional densities $p(\mathbf{x}|\omega_i)$ using only local information at $\mathbf{x}$.   Because its theoretical justification is asymptotic, the information is <u>very</u> local for large n.   Kernel discriminant analysis rather uses local information in a gradual way. The basic kernel density estimate of $p(\mathbf{x}|\omega_i)$ is

$$\hat{p}(\mathbf{x}|\omega_i) = \sum_{j=1}^{n_i} \frac{1}{n_i h^d} K(\frac{\mathbf{x}-\mathbf{x}_{ij}}{h})$$

the sum being over the training points from class ω_i.   Here K(), the kernel, is a pdf such as the multivariate normal density or

$$K_e(\mathbf{x}) = \frac{(d+2)}{2c_d} \max(0, 1-\mathbf{x}^T\mathbf{x})$$

(here c_d is the volume of the unit ball in d dimensions).   Nearby points contribute to $\hat{p}(|\omega_i)$ according to their distance away.   The parameter h controls what is meant by "nearby" and is the most important choice in kernel-based procedures. We saw in section 2.1 that for discrimination it is most important to know $p(|\omega_i)$ accurately in the tails.   This suggests over-smoothing in the centre of the distribution to achieve smooth tails.   One of the best ways to achieve this is to use <u>adaptive</u> kernel estimates of the form

$$\hat{p}(\ |\omega_i) = \frac{1}{n_i h^d} \sum_{j=1}^{n_i} \lambda_i^{-d} K(\frac{x-x_{ij}}{\lambda_i h})$$

where the local bandwidth factors λ_i are determined from a pilot estimate $\tilde{p}(\ )$ by

$$\lambda_i = \{\tilde{p}(x_i)/g\}^{-\frac{1}{2}}$$
$$\ln g = n_i^{-1} \sum \ln \tilde{p}(x_i)$$

This increases λ_i and so smooths more in the tails, where the pilot estimate $\tilde{p}()$ is low.

Theoretical and practical details on the implementation of kernel density estimators are given by Silverman [18].   Their use in discrimination is mentioned by Silverman and discussed in more detail by Coomans & Broeckaert [3] and Hand [9].

2.5 Discrete data

All the methods of this section thus far have assumed that the feature vector **x** has a continuous distribution.   This holds in most of the traditional applications, but frequently not in medical diagnosis, where the results of tests are binary or ordinal.   The theory of the Bayes rule applies as before, but suitable estimators of the $p(\ |\omega_i)$ are needed.

One approach is to model the distribution of **x** by some model from discrete multivariate analysis, such as a log-linear model, McCullagh's [11] ordinal models, Lancaster models or latent class models.   Kernel methods of smoothing were originally proposed in this context by Aitchison and Aitken [1], and discussed in detail by Coomans & Broeckaert [3] and Hand [9].

3   SELECTING THE FEATURE VECTOR

Although the stress in the literature is on designing the classifier <u>given</u> the feature vector **x**, it seems that a good choice of feature vector is essential for satisfactory performance.   At its simplest this means actually finding variables which have discriminatory power between the classes.   Often in biological classification problems it is not the measurements themselves, but combinations of their measuring shape which are essential.   One example is the classification of male and female crabs of two species discussed by Campbell and Mahon[2].   There ratios of width to length proved important, this being discovered by canonical variate analysis on logarithms of the measurements.

It seems that in most PR problems rather little is known about the classes, and lots of measurements are taken in the hope that some of them will be useful.   Since ratios may be important, logarithms and other transformations of the measurements are often added.   As Devijver and Kittler ([4], p.15) comment, "despite the fact that quantity never quite compensates for quality", it does seem to be believed that automatic methods will extract something from a mass of data.

This high dimensionality has several consequences. Firstly, automatic methods are necessary since inspection is nigh impossible. Second, there will be no hope of estimating $p(\ |\omega_i)$ at all accurately, and thirdly, any classifier will take a lot of computing time just in handling the data. Thus some dimensionality reduction is essential. This can take one of two forms:

(a) feature selection. The dimensionality is reduced by selecting a few measurements.

(b) feature extraction.   A small number of combinations of all the original measurements are selected.

In examples such as gaschromatography and spectrophotometry the large number of measurements may be essentially free, so it makes sense to make use of all of them.

3.1 What is a good feature vector?

We would like to choose the feature vector so that the classifier works well. Let $\xi = g(\mathbf{x})$ be the reduced feature vector. Then we would construct a classifier based on ξ and estimate its average error rate $E^*(g)$, then optimize over g to minimize the average error rate. This is theoretically possible, but estimating $E^*(g)$ is so fraught with difficulty and computationally expensive that this procedure is never used. What we need is a good surrogate for low $E^*(g)$.   Often it will be sufficient to achieve good separation in some sense between the training samples or $\hat{p}(\ |\omega_i)$ for the various classes. Many measures of separation have been proposed (Devijver & Kittler, [4], Chapters 5-8, for example).   For example, with $\xi \sim N(\mu_i,\Sigma)$ for $\mathbf{x}$ from class ω_i, we might wish to separate the means as possible whilst keeping the within-class dispersion constant.   This is the aim of canonical variates, which in the current terminology is a feature-extraction method.

3.2 Feature selection and extraction

Given a measure of class separation, this is then maximized numerically. The problem is rather similar to finding good subsets of regressors in multiple regression. The combinatorial explosion makes it impossible to search all subsets, but techniques such as branch-and-bound make it possible to reduce the search to just some of the subsets. Little more can be said in general about feature selection except that some measures of class separation can be computed recursively, and that some are more closely related to average error rate than others.

Feature extraction is usually taken to mean searching for linear transformations g, so $\xi = G\mathbf{x}$ where G is a $d' \times d$ matrix with $d' \ll d$. it is then a continuous optimization problem to find G, and there is some hope of analytical progress. The example of canonical variates shows that an explicit analytical solution may be possible. In fact with certain separability measures and methods of estimating $p(\ |\omega_i)$ considerable progress can be made. For example, consider the Patrick-Fisher [15] probabilistic dependence,

$$J(G) = \sum_{i=1}^{c} \pi_i \sqrt{[\int \{p(G\mathbf{y}|\omega_i) - p(G\mathbf{y})\}^2 d\mathbf{y}]}$$

If $p(\ |\omega_i)$ and hence $p()$ are estimated by a kernel method with a Gaussian kernel, the derivative $J'(G)$ can be written in an explicit way as a function of the training set, and no numerical integration is needed. If we are able to assume that $p(\ |\omega_i)$ is Gaussian this and many other criteria J can be found analytically.

It will be seen that computational considerations dominate discussions of automatic feature selection and extraction.

3.3 The Karhunen-Loève expansion

A more systematic way to proceed with feature extraction is to model the classes by a series expansion, and to truncate the series in a suitable way. Fix a class ω_i and $p(\ |\omega_i)$. Then we seek an expansion

$$\mathbf{x} = \sum_{j=1}^{\infty} y_i \mathbf{u}_j$$

in terms of fixed orthonormal vectors $\mathbf{u}_j$ such that

$$E \left\| \mathbf{x} - \sum_{j=1}^{d} y_j \mathbf{u}_j \right\|^2$$

is minimized. The Karhunen-Loève expansion is the minimizing choice $\mathbf{u}_1,...,\mathbf{u}_d$. Its finite-sample version is the principal components expansion.

Unfortunately the model can differ from class to class. If we are prepared to assume that the classes only differ in location, the K-L model can be applied to the common centred distributions. A feature extractor $\xi = G\mathbf{x}$ is then constructed from those $y_j\mathbf{u}_j$ which maximize some criterion of class separation. The important point here is that because of orthogonality, the d'-dimensional description is made up of the d' best one-dimensional components.

The SIMCA procedure (Wold, [20]) favoured in chemometrics is in essence the finite-sample version of the above. A principal components expansion is constructed for each class, and outliers discarded from the training set. Then a new object is classified as being in class ω_j if it passes a significance test of concordance with the class ω_j model. Multiply classified points become "doubt" decisions, those classified as no class are reported as "not one of these classes". The "in essence" above covers a plethora of operational details.

4   USE OF CONTEXT

Many pattern recognition decisions are not made in isolation. For example in speech or symbol recognition the basic building blocks make up a message, and in classifying landuse in an image, the landuse of blocks of pixels is important.   In such problems it is important to use contextual information.   Several approaches are possible.   One is to regard the statistical PR problem as a small component in a structural PR problem. Another is to enlarge the dataset for each decision to include data from neighbours.   Perhaps the most ambitious form is to enlarge the statistical PR problem to recognizing the whole collection of symbols, landuses (and so on) simultaneously.

Perhaps the distinction is most clearly seen in image analysis.   Consider the landuse problem in more detail.   One way to incorporate contextual information is to classify its pixel not only on the basis of its signal but also on the basis of the signals on the N,E,S and W neighbours of the pixels.   In this approach the decisions on the neighbouring pixels are not used, only the data.   An alternative is to enlarge the problem to choose the most probable a posteriori map of landuse, when the decisions on the neighbours do matter to the decision at a particular pixel, and may dominate that decision.   Details of the methods used are given in Ripley [6].

5   FURTHER READING

Krishnaiah and Kanal [10] provide an overview of the subject, both theory and applications.   More specific introductions to aspects of statistical PR include [3], [4], [7], [8] and [9].

REFERENCES

1. Aitchison, J. & Aitken, C.G.G. (1976).   Multivariate binary discrimination by the kernel method.   Biometrika, **63,** 413-420.

2. Campbell, N.A. & Mahon, R.J. (1974).   A multivariate study of variation in two species of rock crab of genus Leptograspus.   Aust. J. Zool. **22,** 417-425.

3. Coomans, D. & Broeckaert, I. (1986).   Potential Pattern Recognition. Research Studies Press.

4. Devijver, P.A. & Kittler, J.V. (1982).   Pattern Recognition.   A Statistical Approach.   Prentice-Hall.

5. Fisher, R.A. (1936).   The use of multiple measurements in taxonomic problems.   Ann. Eugenics **7,** 179-188.

6. Fix, E. & Hodges, J. (1951).   Discriminatory analysis, nonparametric discrimination : consistency properties.   USAF School of Aviation Medicine project 21-49-004, report 4.

7. Fukunaga, K. (1972).   Introduction to Statistical Pattern Recognition. Academic Press.

8. Hand, D.J. (1981).   Discrimination and Classification.   Wiley

9. Hand, D.J. (1982).   Kernel Discriminant Analysis.   Research Studies Press.

10. Krishnaiah, P.R. & Kanal, L. (1982).   Classification, Pattern Recognition and Reduction of Dimensionality.   Handbook of Statistics **2,** North-Holland.

11. McCullagh, P. (1980).   regression models for ordinal data (with discussion). J. Roy. Statist. Soc. B 42, 109-142.

12. Minsky, M. (1961).   Steps towards artificial intelligence. Proc. IRE 49, 8-30.

13. Minsky, M. & Papert, S. (1969).   Perceptrons: An Introduction to Computational Geometry. MIT Press.

14. Nilsson, N.J. (1965).   Learning Machines : Foundations of Trainable Classifying Systems. McGraw-Hill.

15. Patrick, E.A. & Fisher, F.P. (1969).   Nonparametric feature selection. IEEE Trans. IT-15, 557-584.

16. Ripley, B.D. (1986).   Statistics, images and pattern recognition. Canad. J. Statist. 14, 83-111.

17. Rosenblatt, F. (1962).   Principles of Neurodynamics : Perceptrons and the Theory of Brain Mechanisms.   Spartan Books.

18. Silverman, B.W. (1986).   Density Estimation for Statistics and Data Analysis.   Chapman and Hall.

19. Tukey, P.A. & Tukey, J.W. (1981).   Graphical display of data in 3 or more dimensions. In Interpreting Multivariate Data, V. Barnett (ed), Wiley, pp.189-275.

20. Wold, S. (1978).   Pattern recognition by means of disjoint principal component models. Pattern Recgn. 8, 127-137.